纺织服装高等教育“十四五”部委级规划教材

Fibres, Yarns and Fabrics

纺织英语

（第四版）

编　　著：卓乃坚 Zhuo Naijian

英文审读：西蒙 C.哈罗克博士 Dr. Simon C. Harlock

東華大學出版社

·上海·

内容提要

本书用英语介绍纤维、纱线、织物形成及染色、印花和整理等过程，涉及上千条常用纺织术语。为了方便读者理解，本书使用了不少插图，并且每章后附有练习题和阅读材料。另外，书后还附有中文参考译文和便于查阅的词汇表。

本书可以用作纺织专业学生的专业英语教材，也可以作为国际贸易专业学生了解纺织的双语教材，还可以作为有关纺织工作者及纺织品外贸工作者有益的参考读物。

图书在版编目（CIP）数据

纺织英语 / 卓乃坚编著. —4 版. —上海：东华大学出版社，2022.5

ISBN 978-7-5669-2054-6

Ⅰ.①纺… Ⅱ.①卓… Ⅲ.①纺织—英语—教材 Ⅳ.①TS1

中国版本图书馆 CIP 数据核字（2022）第 072827 号

责任编辑：张　静
封面设计：魏依东

出　　版：东华大学出版社（上海市延安西路 1882 号，200051）
本社网址：http://dhupress.dhu.edu.cn
天猫旗舰店：http://dhdx.tmall.com
营销中心：021-62193056　62373056　62379558
印　　刷：上海四维数字图文有限公司
开　　本：787 mm×960 mm　1/16　印张 13
字　　数：324 千字
版　　次：2022 年 5 月第 4 版
印　　次：2022 年 5 月第 1 次印刷
书　　号：ISBN 978-7-5669-2054-6
定　　价：59.00 元

Reasonable efforts have been made to publish reliable data and information, but the author and the publisher cannot assume responsibility for the validity of all materials. Neither the author nor the publisher, nor anyone else associated with this publication, shall be liable for any loss, damage or liabilities directly or indirectly caused or alleged to be caused by this book. 为出版可靠的数据及信息,作者和出版商已作了合理的努力,但他们不可能对所有材料的有效性负责任。作者、出版商或任何与本出版物关联的人不对直接、间接或声称由本书造成的任何损失、损害或债务承担法律责任。

Neither this book nor any part of it may be reproduced or transmitted in any form or by any means, electronic or mechanical, including photocopying, microfilming and recording, or by any information storage or retrieval system, without written permission from the author or the publisher. 未经作者或出版商书面许可,本书或其任何部分不可以任何形式或任何电子的或机械的手段,包括复印、缩微拍摄、录制,或通过任何信息储存或检索系统,进行复制或传递。

Preface to the Forth Edition

The COVID-19 epidemic is still keeping impact on every aspect of the world, but however, the world textile industries are also keeping development, the themes of which are innovation and sustainability. Around these themes, we update all reading materials for the present edition in the hope that after getting the fundamental knowledge about fibres, yarns and fabrics from the text of this book, our readers could get fresh knowledge about the current trends of the textile abroad.

Those elaborately selected reading materials are excerpted from recent foreign English textile magazines, and many of them have digital issues which could be reachable from the relative websites, and this would greatly facilitate our readers to get further learning when they need. In order to make our readers easily understand, all reading materials are attached with necessary reference tips.

Special thanks are given to Mr. Zhuo Shufan for all work he has done in retrieving and preparing the reading materials adopted in this edition.

The author

April 2022

前　言

新冠疫情仍在持续影响着世界的方方面面,然而世界纺织业也在不断发展,创新和可持续性是这个发展的主题。围绕这些主题,我们在本书四版中更新了所有阅读材料,目的就是使读者在了解了本书课文中的关于纤维、纱线和织物的基本知识后对当今国外纺织动态有新的了解。

这些精心挑选的阅读材料摘自国外近期的英语纺织杂志,其中不少杂志在相关网站上可以查阅到电子版,这大大方便了读者在需要时做进一步的了解。为了方便读者理解,本书所有阅读材料都加上了必要的参考提示。

特别感谢卓书帆为检索和准备本版阅读材料所做的所有工作。

作者

2022 年 4 月

Preface to the Third Edition

The textile industry is one of the oldest industries in the world. Although some people think that it is a "sunset" industry, it is clear that no one can live without the products from this industry. Textile products are playing an essential role not only in our everyday life but also in the fields of medical treatment, engineering, and aerospace technology.

It is a commonly held belief that a country should develop its industries according to its competitive advantages and it is the same for the textile industry. The textile industry in some developed countries has tended towards capital-intensive industry whereas in developing countries the industry is labour-intensive one. Such a tendency, no doubt, has greatly promoted the development of textile international trade.

As the globalization continues, communication and information are becoming more and more important for the international textile business. Although in today's world there are many means of communication and ways to search for information, language is still of the utmost importance.

English is the principal language in international communication and for anyone involved in the textile business, it is almost essential to know many specialized textile terms in English, especially if he or she intends to be engaged in international business, or is seeking outside information to develop textile industry.

Therefore the purpose of this book is to provide readers with both a fundamental insight into fibres, yarns and fabrics, and, in doing so, introduce them to many specialized textile terms in both English and Chinese.

It is envisaged that this book will serve as a suitable textbook for both students majoring in textiles as well as students majoring in international trade who are intending to become involved in textile manufacture.

What's New in this edition? Like the second edition, reading materials excerpted from some well-known English textile magazines, published in U. K.,

U. S. A. etc. , are provided for each chapter. All reading materials in the last edition are replaced by those from the latest magazines and some useful terms in the old reading materials are kept and incorporated into the text part of the relative chapters, and thereby the author wishes to present to readers the newest information about the textile world. If this book is used as a textbook, those reading materials could be used as exercises for students to make translations.

Acknowledgement. The author would like to express his profound gratitude and sincere appreciation to Dr. Simon C. Harlock for his important contribution to this book, and the author also wish to take this opportunity to express his immense gratitude and deep reverence to all who enable him to grasp the solid knowledge in English and textiles, including Dr. G. A. V. Leaf, his (ex) Ph. D. supervisor in the University of Leeds; Jin Yuyan, his (ex) tutor of specialized English in East China Textile Institute of Science and Technology (now called Donghua University); Shen Erkang, his (ex) English teacher in middle school.

Special thanks are also extended to Mr. Zhuo Shufan for his contributions in preparing illustrations, vocabulary list and retrieving materials for this book.

The author

December 2016

第三版前言

纺织业是世界上最古老的工业之一,尽管有人认为它是一个夕阳产业,但不容置疑,人们的生活离不开它的产品。纺织产品不仅在我们日常生活中,而且在医疗、工程及航天技术领域中,正发挥着必不可少的作用。

通常,人们都认为,一国应该根据自己的竞争优势发展自己的产业,纺织业也是如此。某些发达国家的纺织业已趋向成为资本密集型产业,而发展中国家的纺织业仍然是劳动力密集型的。无疑,这种趋势大大促进了纺织国际贸易的发展。

随着全球化的持续,交流和信息对于国际纺织商务越来越重要。尽管在当今的世界,交流的手段和信息搜索的方式很多,语言仍然是至关重要的。

英语是国际交往中的主要语言,对于涉足纺织商务的每个人,了解很多的英语纺织专业术语几乎是必要的,尤其如果他/她打算参与国际商务,或寻求外部信息以发展纺织业。因此本书的目的就是让读者对纤维、纱线和织物有一个基本了解,同时向他们介绍很多的英文和中文的纺织专业术语。

本书设想可以作为纺织专业和预期将涉足纺织制造的国际贸易专业的学生的教材。

本版有何新处? 如同第二版,本版为每一章提供了一些摘自英美等国出版的知名英文纺织杂志的阅读材料。上一版中的所有阅读材料被那些摘自近期的杂志中的文章所替代,而老的阅读材料中某些有用的术语得以保留并融入相应各章的课文中,以此作者希望能为读者提供最新的纺织界信息。如果本书作为教材,这些阅读材料可以用作学生的翻译练习。

致谢。 作者对西蒙 C. 哈罗克博士对本书所做的重要贡献深表谢意,作者还想借此机会对所有使他掌握英语和纺织方面的坚实知识的人表达无尽的谢意和深深的敬仰,其中有:G. A. V. 里夫博士,作者在利兹大学攻读博士时的导师;金玉燕,作者在华东纺织工学院(现为东华大学)时的专业英语指导老师;沈尔康,作者在中学时期的英语老师。另外,特别感谢卓书帆为制备本书的插图和词汇表以及检索本书所用材料所做的贡献。

作者

2016 年 12 月

Contents | 目录

CHAPTER 1

TEXTILE FIBRES

Fibres are the basic elements of textiles. Generally speaking, materials with diameters ranging from several microns to tens of microns and with lengths being many times of their thickness can be considered to be fibres. Among them, those longer than tens of millimetres with sufficient strength and flexibility can be classified as textile fibres, which can be used to produce yarns, cords or fabrics.

1 TYPES OF TEXTILE FIBRES

There are many types of textile fibres. However all may be classified as either natural fibres or man-made fibres.

1.1 NATURAL FIBRES

Natural fibres include plant or vegetable fibres, animal fibres and mineral fibres.

In terms of popularity, cotton is the most commonly used plant fibre, followed by linen (flax) and ramie. Flax fibres are commonly used, but since the fibre length of flax is fairly short (25 ~ 40 mm), flax fibres have traditionally been blended with cotton or polyester. Ramie, the so-called "China grass", is a durable bast fibre with a silky lustre. It is extremely absorbent but the fabrics made from it crease and wrinkle easily, so ramie is often blended with synthetic fibres.

Animal fibres either come from the animal's hair, for example, wool, cashmere, mohair, camel hair and rabbit hair, etc., or from the animal gland secretion, such as mulberry silk and tussah.

The most commonly known natural mineral fibre is asbestos, which is an inorganic fibre with very good flame resistance but is also dangerous to health and, therefore, is not used now.

1.2 MAN-MADE FIBRES

Man-made fibres can be classified as either organic or inorganic fibres. The former can be sub-classified into two types: one type includes those made by transformation of natural polymers to produce regenerated fibres as they are sometimes called, and the other type is made from synthetic polymers to produce synthetic filaments or fibres.

Commonly used regenerated fibres are Cupro fibres (CUP, cellulose fibres obtained by the cuprammonium process) and Viscose (CV, cellulose fibres obtained by the viscose process. Both Cupro and Viscose can be called rayon). Acetate (CA, cellulose acetate fibres in which less than 92%, but at least 74%, of the hydroxyl groups are acetylated.) and triacetate (CTA, cellulose acetate fibres in which at least 92% of the hydroxyl groups are acetylated.) are other types of regenerated fibres. Lyocell (CLY), Modal (CMD) and Tencel are now popular regenerated cellulose fibres, which were developed to meet the demand for environmental consideration in their production.

Nowadays regenerated protein fibres are also becoming popular. Among these are soyabean fibres, milk fibres and Chitosan fibres. Regenerated protein fibres are particularly suited for medical applications.

Synthetic fibres used in textiles are generally made from coal, petroleum or natural gas, from which the monomers are polymerized through different chemical reactions to become high molecular polymers with relatively simple chemical structures, which can be melted or dissolved in suitable solvents. Commonly used synthetic fibres are polyester (PES), polyamide (PA) or Nylon, polyethylene (PE), acrylic (PAN), modacrylic (MAC), polypropylene (PP) and polyurethane (PU). The aromatic polyesters such as polytrimethylene terephthalate (PTT), polyethylene terephthalate (PET) and polybutylene terephthalate (PBT) are also becoming popular. In addition to these, many synthetic fibres with special properties have been developed, of which Nomex, Kevlar and Spectra fibres are well known. Both Nomex and Kevlar are the registered brand names of the Dupont Company. Nomex is a meta-aramid fibre with an excellent flame retardant property and Kevlar can be used to make bullet-proof vests because of its extraordinary strength. Spectra fibre is made from polyethylene, with ultra-high molecular weight, and is considered to be one of the strongest and lightest fibres in the world. It is particularly suited for armour, aerospace and high-performance sports goods. Research is still going on. The research on nano fibres is one of the hottest topics in this field and in order to ensure that nanoparticles are safe for man and the environment, a new field of science called "nanotoxicology" is derived, which currently looks at developing test methods for investigating and evaluating the interaction between nanoparticles, man and environment.

Commonly used inorganic man-made fibres are carbon fibres, ceramic fibres, glass fibres and metal fibres. They are mostly used for some special purposes in order to perform some special functions.

2 PROPERTIES OF TEXTILE FIBRES

Much research has been conducted into the properties of textile fibres. These include sorption properties to find whether a particular fibre is hydrophilic, hydrophobic, hygroscopic, oleophilic, or oleophobic, and other properties, such as tenacity, elastic recovery, abrasion resistance, flexibility, creep properties, combustibility, chemical properties and resistance to biological organisms, etc. Figure 1.1 presents a brief summary of the performance characteristics of some common textile fibres.

Generally speaking, protein fibres have higher resilience and they are hygroscopic or hydrophilic fibres, whose mechanical properties change as they absorb moisture. Alkalis impair their mechanical properties and ultraviolet light may cause them to yellowing and weaken. The actual properties of different protein fibres differ according to their particular morphological and chemical structures. For example, wool has a scaly surface, which makes it prone to felting unless treated to prevent it. In contrast, silk has a smooth surface, which imparts a shiny lustre to it.

Cellulosic fibres are also hydrophilic, and their mechanical properties will also change after moisture absorption. Compared to natural protein fibres, they have lower resilience and much better resistance to alkaline degradation. Among them, ramie has excellent tenacity and very good resistance to UV light. Their specific morphologies and chemical structures also affect their properties. An individual cotton fibre is convoluted, like a deflated hose-cotton has lower thermal insulation because the lumen in most cotton fibres collapses as the fibre dries out after growing. The chemical

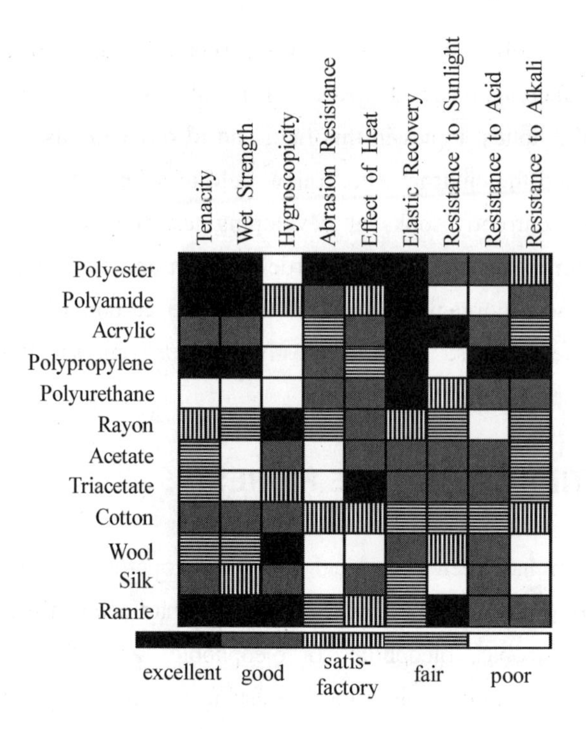

Fig. 1.1 Performance ratings of prominent textile fibres

component of the viscose fibres is similar to that of the cotton, which gives them similar characteristics. However, since the degrees of polymerization and crystallization in viscose are lower than those of cotton, viscose has a better hydroscopic property but poorer tensile strength, especially in the wet state.

Synthetic fibres generally have lower moisture regain, and most of them are oleophilic but hydrophobic. Unlike the protein and cellulosic fibres, synthetic fibres have good resistance to moths, mildew and fungi. The actual properties of synthetic fibres will depend on their molecular length, chemical composition, arrangement of polymers, bonds between the molecules and the shape of their cross sections, etc. For example, the more the amorphous regions and the more the H-bonds or polar groups that are present within the structure, the more hydrophilic the fibre will be; the more the molecules are orientated in the axial direction of the fibre, the higher the fibre's tenacity will be. To increase the orientation, most synthetic filaments were stretched or drawn during their manufacture. The properties of the synthetic fibres are highly dependent on their chemical composition. For example, polyester has good tenacity due to its higher crystallinity, poor hydrophilicity due to a lack of hydrophilic groups, good resistance to acid but less good to alkalis due to its chemical composition.

3 THE QUALITY OF FIBRES

Attention must be paid to fibre quality, because this critically affects the quality of yarns and fabrics made from them. The quality of fibres can be considered from two perspectives, *viz.* apparent quality and inherent quality.

Fibres that are stuck together during their manufacture will affect the apparent qualities of man-made fibres, and furthermore, faults in their appearance would also affect the inherent quality. Another example is the sulphur spots that can occur in viscose fibres due to insufficient desulphuration.

The inherent qualities are principally the mechanical and chemical qualities, which might affect the later processing or the end-use of fibres. Measurements of breaking strength, elongation at break, fibre length variation, regularity of fibre fineness, proportion of over-length fibres, crimp frequency and moisture regain, etc. need to be made in order to evaluate fibre quality. For some fibres, further tests may be required such as wet tenacity, loop strength and residual sulphur content for viscose, dye-uptake rate for acrylic and boiling shrinkage for polyester.

For natural fibres, tensile tests are performed to determine their breaking strengths and extensibility, from which the variabilities in strength or elongation might be calculated. The fineness of the fibre specimens are usually measured to find their mean values and variation coefficients. Any impurities, such as vegetable content in cotton or oil in wool, need to be checked, and they are important factors in the evaluation of fibre quality.

Before testing, textile fibres should be conditioned to bring the testing material into moisture equilibrium with standard atmosphere for testing. Testing on textile fibres is conducted in standard atmospheric conditions which are a relative humidity of 65% ± 2% and a temperature of 20℃ ± 2℃ in the air at local atmospheric pressure. For some materials, such as polyester and acrylic, which are known to be relatively unaffected by changes in relative humidity, the tolerance in relative humidity can be extended to ± 5%. For tests on yarns or fabrics, the same conditions should also be strictly followed, especially for those sensitive to humidity.

Words and Phrases

fibre [ˈfaɪbə]	纤维
textile	纺织品
micron [ˈmaɪkrɒn]	微米
natural fibre	天然纤维
man-made fibre	化学纤维
synthetic [sɪnˈθetɪk] fibre	合成纤维
plant fibre	植物纤维
vegetable fibre	植物纤维
animal fibre	动物纤维
mineral [ˈmɪnərəl] fibre	矿物纤维
cotton	棉
linen [ˈlɪnɪn]	亚麻织物或纱线
ramie [ˈræmɪ, ˈreɪ-]	苎麻
flax [flæks]	亚麻
polyester [ˌpɒlɪˈestə] (PES)	涤纶/聚酯
bast [ˈbæst] fibre	韧皮纤维
wool	羊毛
cashmere [kæʃˈmɪə]	羊绒/开司米
mohair [ˈməʊheə]	马海毛
camel [ˈkæməl] hair	驼毛
rabbit hair	兔毛
animal gland [glænd] secretion [sɪˈkriːʃən]	动物腺分泌液
mulberry [ˈmʌlbərɪ] silk	桑蚕丝
tussah [ˈtʌsə]	柞蚕丝
asbestos [æzˈbestɒs]	石棉
organic [ɔːˈgænɪk] fibre	有机纤维
inorganic [ˌɪnɔːˈgænɪk] fibre	无机纤维
polymer	聚合物,聚合体
regenerated fibre	再生纤维

filament [ˈfɪləmənt]	长丝
Cupro [ˈkjuːprəʊ] fibre (CUP)	铜氨纤维
cellulose [ˈseljʊləʊs] fibre	纤维素纤维
cuprammonium [ˌkjuːprəˈməʊnɪəm] process	铜氨法加工
viscose [ˈvɪskəʊs] (CV)	黏胶
viscose process	黏胶法加工
rayon [ˈreɪɒn]	(铜氨法或黏胶法加工而成的)人造再生纤维素纤维
acetate [ˈæsɪteɪt] (CA)	醋酯(纤维)
hydroxyl [haɪˈdrɒksɪl] groups	羟基
acetylated [ˌæsɪtɪˈleɪtɪd]	乙酰化的
triacetate [traɪˈæsɪˌteɪt] (CTA)	三醋酯(纤维)
Lyocell [laɪəˈsel] (CLY)	绿塞尔(纤维)
Modal [ˈməʊdəl] (CMD)	莫代尔(纤维)
Tencel [ˈtensɪl]	天丝
regenerated protein fibre	再生蛋白质纤维
soybean [ˈsɔɪbiːn] fibre	大豆纤维
milk fibre	牛奶纤维
chitosan [ˈkaɪtəsæn] fibre	甲壳素纤维
monomer [ˈmɒnəmə]	单体
polymerise [ˈpɒlɪməraɪz, pəˈlɪ-]	聚合
high molecular polymer	高分子聚合物
polyamide [pɒlɪˈæmaɪd] (PA)	锦纶/聚酰胺
nylon	尼龙
polyethylene [ˌpɒlɪˈeθɪliːn] (PE)	聚乙烯
acrylic [əˈkrɪlɪk] (PAN)	腈纶/聚丙烯腈
modacrylic [ˌmɒdəˈkrɪlɪk] (MAC)	变性聚丙烯腈
polypropylene [ˌpɒlɪˈprəʊpliːn] (PP)	丙纶/聚丙烯
polyurethane [ˌpɒlɪˈjʊərɪθeɪn] (PU)	氨纶/聚氨基甲酸酯
aromatic [ˌærəʊˈmætɪk]	芳族的
polytrimethylene [ˈpɒlɪtraɪˈmeθɪliːn]	聚对苯二甲酸丙二酯

terephthalate [ˈteˌrefˈθæleɪt] (PTT)	
polyethylene terephthalate (PET)	聚对苯二甲酸乙二酯
polybutylene [ˌpɒlɪˈbjuːtɪliːn] terephthalate (PBT)	聚对苯二甲酸丁二酯
Nomex [ˈnəʊmeks] fibre	诺梅克斯纤维(聚间苯二甲酰间苯二胺纤维,杜邦公司品牌)
Kevlar [ˈkevlɑ] fibre	凯芙拉纤维(杜邦公司品牌)
Spectra [ˈspektrə] fibre	斯佩克特纤维(霍尼韦尔公司品牌)
Dupont Company	杜邦公司
meta-aramid fibre	间位芳族聚酰胺纤维
flame retardant [rɪˈtɑːdənt]	阻燃的
bullet-proof vest	防弹背心
molecular weight	相对分子质量
armour [ˈɑːmə]	盔甲
nano [ˈnænəʊ, ˈneɪnəʊ] fibre	纳米纤维
nanoparticle [ˈnænəʊpɑːtɪkl]	纳米粒子
nanotoxicology [ˈnænətɒksɪˈkɒlədʒɪ]	纳米毒理学
carbon fibre	碳纤维
ceramic [sɪˈræmɪk] fibre	陶瓷纤维
glass fibre	玻璃纤维
metal fibre	金属纤维
sorption properties	吸着性质
hydrophilic [ˌhaɪdrəʊˈfɪlɪk]	亲水的
hydrophobic [ˌhaɪdrəʊˈfəʊbɪk]	疏水的
hygroscopic [ˌhaɪgrəʊˈskɒpɪk]	吸湿的
oleophilic [ˌəʊlɪəʊˈfɪlɪk]	亲油的
oleophobic [ˌəʊlɪəʊˈfəʊbɪk]	疏油的
tenacity [tɪˈnæsɪtɪ]	强度
elastic recovery	弹性回复
abrasion [əˈbreɪʒən] resistance	耐磨性
flexibility [ˌfleksəˈbɪlɪtɪ]	韧性
creep property	蠕变性质

combustibility [kəmˌbʌstəˈbɪlɪtɪ]	可燃性
biological organism	生物体
resilience [rɪˈzɪlɪəns]	回弹性
mechanical properties	机械性质
moisture	水分
alkali [ˈælkəlaɪ]	碱
ultraviolet [ˌʌltrəˈvaɪəlɪt] (UV) light	紫外线光
morphological [ˌmɔːfəˈlɒdʒɪkəl]	形态(学)的
scaly [ˈskeɪlɪ] surface	鳞片层表面
prone [prəʊn] to	易于……的
felting [ˈfeltɪŋ]	毡化
lustre [ˈlʌstə]	光泽
alkaline [ˈælkəlaɪn] degradation [ˌdegrəˈdeɪʃən]	碱性降解
morphology [mɔːˈfɒlədʒɪ]	形态学,表面形状
lumen [ˈljuːmɪn]	内腔
polymerisation [ˌpɒlɪməraɪˈzeɪʃən, pəˈlɪ-]	聚合化
crystallization [ˈkrɪstəlaɪˈzeɪʃən]	结晶化
in the wet state	在湿态
moth [mɒθ]	蛀虫
mildew [ˈmɪldjuː]	发霉
fungi [ˈfʌndʒaɪ, ˈfʌŋgaɪ]	真菌
molecular length	分子长度
chemical composition [kɒmpəˈzʃən]	化学成分
cross section	横截面
amorphous [əˈmɔːfəs] regions	无定形区
H-bond	氢键
polar [ˈpəʊlə] group	极性基团
orientation	定向
crystallinity [ˌkrɪstəˈlɪnɪtɪ]	结晶度
apparent [əˈpærənt] quality	外观品质
inherent [ɪnˈhɪərənt] quality	内在品质

sulphur [ˈsʌlfə] spot	硫斑
desulphuration [ˌdiːsʌlfəˈreɪʃən]	脱硫
breaking strength	断裂强度
elongation [ˌiːlɒŋˈgeɪʃən] at break	断裂伸长
fibre length variation [ˌveərɪˈeɪʃən]	纤维长度偏差
regularity [ˌregjʊˈlærɪtɪ] of fibre fineness	纤维细度均匀度
proportion of over-length fibres	倍长率
crimp frequency	卷曲频率
moisture regain [rɪˈgeɪn]	回潮率
wet tenacity	湿强
loop strength	钩接强度
residual [rɪˈzɪdjʊəl] sulphur content	残硫量
dye-uptake rate	上色率
boiling shrinkage [ˈʃrɪŋkɪdʒ]	沸水收缩率
extensibility	延伸性
variabilities [ˌveərɪəˈbɪlɪtɪs] in strength or elongation	断裂强度或伸长的偏差
specimen	样本,抽样
mean and coefficient [ˌkəʊɪˈfɪʃənt] variation	平均值及变异系数
impurity [ɪmˈpjʊərɪtɪ]	杂质
vegetable content	杂草屑
moisture equilibrium [ˌiːkwɪˈlɪbrɪəm]	湿平衡
relative humidity [hjuːˈmɪdɪtɪ]	相对湿度
atmospheric [ˌætməsˈferɪk] pressure	大气压
tolerance [ˈtɒlərəns]	容差

Exercises

1. Which of the following fibres do you think are suitable for underwear? (　　)

   a) Polyester　　b) Cotton

   c) Polypropylene　　d) Rayon

2. In your answer to Question 1 which of the following supports your decision? ( )

   a) Resistant to sunlight b) Hygroscopic

   c) Oleophilic d) Hydrophobic

3. Compared with nylon, polyester is a better material for curtains because it ________. ( )

   a) has a higher tenacity b) is acid resistant

   c) is resistant to sunlight d) is hydrophobic

4. Which of the following fibres are cellulosic fibres? ( )

   a) Wool b) Lyocell

   c) Cotton d) Acetate

5. Which of the following are synthetic fibres? ( )

   a) Triacetate b) Nomex

   c) Tencel d) Nylon

6. The main difference between acetate and triacetate fibre is that ________. ( )

   a) acetate fibre is a cellulosic fibre but triacetate fibre is not

   b) the proportions of acetylated hydroxyl groups in each fibre type are different

   c) triacetate has better hydrophilicity

   d) triacetate is composed of three acetate molecules

7. As far as the fibre property is concerned, nylon has poor ________. ( )

   a) resistance to sunlight b) wet strength

   c) resistance to acid d) resistance to alkali

8. Which is (are) the main reason(s) why one synthetic fibre has better hydrophilicity than another synthetic fibre? ( )

   a) It has more amorphous regions.

   b) It has more H-bonds or polar groups.

   c) It has a smoother surface.

   d) It has a higher molecular weight.

9. In order to establish the inherent quality of cotton fibres, ________ is usually tested. ( )

a) breaking strength
b) extensibility
c) fineness
d) wet tenacity

10. Testing the inherent quality of cotton fibres should be conducted ________. (   )
    a) after the specimens have been properly conditioned
    b) under a normal room temperature and humidity
    c) under a relative humidity of 65%±2% and a temperature of 20℃±2℃
    d) under a relative humidity of 60%±5% and a temperature of 20℃±5℃

Reading Materials

Toray Commercializes Ecodear N510 100% Plant-based Nylon Fiber

Toray Industries has developed a nylon 510 (N510) fiber that incorporates 100% bio-based synthetic polymer content as defined under section 3.1.5 of ISO 16620-1: 2015, the international standard for the bio-based content of plastics. Ecodear N510 will be the first 100% plant-based nylon fiber in Toray's Ecodear lineup.

While primarily for sports and outdoor fabrics, potential applications also include lightweights, cut-and-sew fabrics and innerwear lace materials.

Toray plans to begin Ecodear N510 textiles sales for fall/winter 2023. Initial production volume is expected to be 200,000m by the end of March 2023, growing to 600,000m in March 2026. Ecodear N510 fiber sales are targeted for fall/winter 2024, with an expectation of a monthly supply of 3 metric tons monthly in the year ending March 2024.

— excerpted from *International Fiber Journal*, Issue 1 2022, page 11

【参考提示】

1. *International Fiber Journal*,美国 Inda Media 出版的双月刊。可以看到文章中 fiber 为美式拼法。

2. Toray Industries,日本东丽工业株式会社,Ecodear 是该公司生物基尼龙纤维的注册商标。在做翻译练习时,外文资料中的专有名词除非有普遍接受的中文译名,不妨直接借用原文,如"...Ecodear N510 全植物基尼龙纤维"。

3. plant-based,植物基的;bio-based,生物基的。

4. section 3.1.5 of ISO 16620-1:2015,编号 16620-1:2015 国际标准下第 3.2.5 节。严格来说,应该是 Section 3.1.5,因为专有名词首字母应该大写。

5. light-weight fabric,薄型织物。

6. innerwear,即 underwear,内衣。

7. ... with an expectation of a monthly supply of 3 metric tons monthly in the year ending March 2024,……预计在截至 2024 年 3 月的(过去)一年中,每月供应量为 3 公吨。英语表述中第二个 monthly 似乎多余。

Zoltek Invests in Carbon Fiber Capacity

Zoltek Companies Inc. — with production facilities in the United States, Hungary and Mexico; and a U. S. subsidiary of Japan-based Tory Industries Inc. — has announced a capacity expansion in its large-tow carbon fiber production in 2026. Large-tow carbon fiber is fiber with more than 40,000 filaments and is often used in industrial applications such as wind turbine blades. The $ 130 million investment will be used to increase capacity at a Zoltek facility in Jalisco, Mexico, by approximately 54 percent to more than 20,000 tons per year. Once the new capacity is up and running, combined annual production of Zoltek's Mexico and Hungary plants will be 35,000 metric tons.

— excerpted from *Textile World*, November/ December 2021, page 6

【参考提示】

1. *Textile World*,美国 Textile Industries Media Group 出版的双月刊。
2. Zoltek,美国知名的碳纤维制造公司,国内有译为"卓尔泰克"的。
3. large-tow carbon fiber,大丝束碳纤维。
4. ... more than 40,000 filaments,这里的 filament 指"单丝纤维""原丝"。
5. wind turbine blades,风力(涡轮)发电机叶片。

CHAPTER 2

YARNS

Yarns are the fundamental elements for knitted or woven fabrics. A yarn may be produced from staple fibres or continuous filaments. The choice of fibre or filament and the structure of the yarn will critically affect the property of fabrics subsequently produced from them.

1 FILAMENT YARNS

Filaments are actually fibres of very great length and consequently are normally called "continuous" filaments. Except for silk, which is a natural filament, most filaments used in the textile industry are man-made filaments. Filament yarns consisting of only a single filament are called monofilaments and those consisting of two or more filaments are called multifilaments. Multifilaments are commonly used in knitting or weaving because they are softer and more flexible.

There are two ways to spin man-made filaments or fibres — melt spinning and solvent spinning. Polymers, such as polyester, polyamide or polypropylene, which melt before decomposing, are spun through melt spinning. Polymer chips are melted, and then the molten polymer is extruded through spinneret holes and solidifies as the filaments emerge into the surrounding air (or water). Most of these polymer chips are made from petroleum oil, and therefore any fluctuation in the oil price in the international market would cause a fluctuation in the prices of the relative man-made fibres.

Solvent spinning is used for polymers which have no clearly defined melting temperature or which decompose or carbonize before reaching their melting temperatures. Solvent spinning is further classified as either dry spinning or wet spinning. In solvent spinning, the spinning solution should be made first, and then, in the case of wet spinning, the polymer solution extruded from the spinneret goes through a kind of liquor in which the filaments solidify and, in the case of dry spinning, through hot air (which evaporates the solvent). Since, in dry spinning, more special measures have to be taken to control the environmental pollution, it generally costs more than wet spinning. Acetate and triacetate filaments and fibres are made through dry spinning, while acrylic and viscose through wet spinning.

The filaments initially formed from the spinneret cannot normally be used in textile applications. The filaments are invariably stretched or "drawn" to orientate the molecular chains and increase the strength and extensibility of the filaments. Furthermore, bundling, washing, oiling, crimping and heat setting processes are performed to make the fibres much stronger and suitable for textile end-uses.

Man-made filaments might shine with lustre, but if delustrant, e. g. titanium dioxide (also called titanox), is added, dull or semi-dull fibre can be made depending on the amount of delustrant added.

With the development of manufacturing technology, and with the improvement in the design of the spinneret or the shape of the holes in the spinneret, various bicomponent fibres, micro fibres and unconventionally shaped fibres are now routinely spun.

Of course, in the manufacture of continuous filaments, attention should be paid to the quality, and before the filament can be packed and dispatched to the knitting or weaving mill, drawing, twisting, washing, heat setting and winding might be specified. The last process is to make the filaments into suitable packages required for subsequent processing by the customers.

Filaments may be textured to entangle the filaments and make them bulkier and more extensible. This may be achieved by false-twisting the filaments and heat setting the twist, and then de-twisting the filaments. Another common method is the air-texturing technique in which filaments are overfed into a turbulent stream of air.

The filaments made in the afore-mentioned way may be cut or stretched broken into fibres called "staple" which refers to their shorter length. When man-made filaments are cut or stretched broken into staple fibres, the fibre length should be controlled to produce the required fibre length distribution. Generally the fibres with a length below 40 millimetres are referred to as short staple or cotton length fibres; and fibres with a length between 60 millimetres to 70 millimetres are referred to as long staple or wool length fibres. Those with a length between 40 to 60 millimetres are medium length fibres.

2 STAPLE YARNS

Staple yarns, or spun yarns, are made from the staple fibres, which include natural fibres such as cotton, wool or linen, or man-made fibres such as polyester, polyamide or acrylic, or mixtures which are referred to as "blended" yarns.

2.1 RING SPUN YARNS

In order to combine staple fibres into a yarn with enough coherence to ensure sufficient strength and extensibility for subsequent processing, the fibres need to be orientated as undirectionally as possible and twisted. The way to form a yarn by adding twist is called yarn spinning. Probably the most commonly used spinning technology is ring spinning, the detailed process of which, to some extent, depends on the type of fibres involved and the final specification for the yarn. A typical ring spinning sequence for cotton yarns is as follows:

Pre-treatment

↓

carding→(combing)→drafting→roving→spinning

Firstly, staple fibres, which arrive at the spinning mill in bales, are opened, cleaned and blended in order to produce yarns of uniform quality and to a specific price. If blended yarns are being produced, different types of fibres can be mixed at this stage according to a pre-determined ratio, usually by weight. Then, the fibres are separated and partially aligned by wire-toothed or saw-toothed cloth covering a set of rotating cylinders on a carding machine, which forms the fibres into a card sliver (see Fig. 2.1).

Figure 2.1 shows an example about how the lap after the opening, cleaning and mixing processes is converted into card slivers. The lap wound on the lap rod is mounted on the lap holder and the rotation of the lap drum makes the lap unwind. When the lap reaches the feeding plate, the feeding roller pushes the lap towards the licker-in (or taker-in as it is also referred to) which is covered with metallic saw-toothed cloth. The licker-in rotates at high speed and pulls out small tufts of fibres from the lap. The centrifugal force, together with the function of the mote knife, causes most of the impurities and very short fibres in the lap to separate and drop through the grids in the under-casing beneath the licker-in. The cylinder is also covered with metallic saw-toothed cloth and rotates at very high speed. The interaction between the teeth on the cylinder and the teeth on the revolving flats above the cylinder opens and separates the tufts of fibres. This interaction is called carding. At the same time, most of the remaining impurities and short fibres are removed. Some of them drop through the grids in the cylinder under casing; some form strips of fibres on the flats. These flat strips are removed by a flat stripping comb and a flat stripping brush. The single fibres after carding are transferred from the cylinder to a slower rotating doffer covered also with metallic saw-toothed cloth. Stripping rollers remove the fibres in the form of a card web from the doffer. Two nip rollers crush any remaining impurities before the web enters a trumpet that condenses it into a card sliver. The sliver is then coiled into a can.

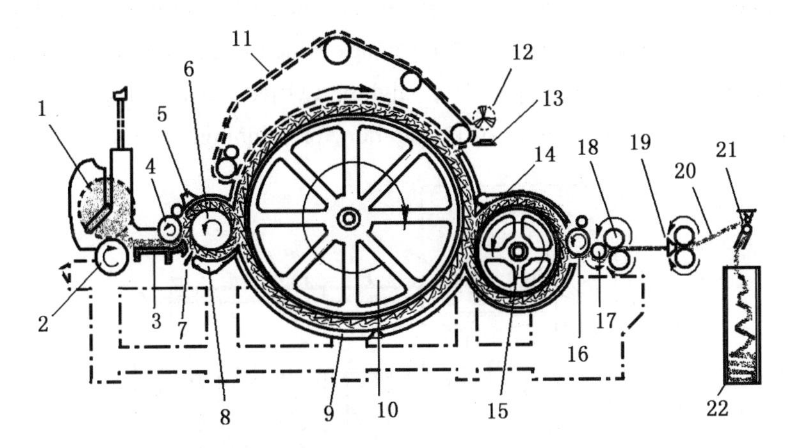

Fig. 2.1 Carding machine

1—lap 2—lap drum 3—feeding plate 4—feeding roller 5—suction cover 6—licker-in 7—mote knife 8—licker-in under casing 9—cylinder under casing 10—cylinder 11—flats 12—flat stripping brush 13—flat stripping comb 14—suction cover 15—doffer 16—stripping roller 17—transferring roller 18—nip roller 19—trumpet 20—card sliver 21—can coiler 22—sliver can

The card described is characteristic of a cotton or short staple carding machine. Modern versions of this machine have the fibres fed through a chute to the licker-in rather than from a lap. Woollen and worsted cards work on a similar principal except that the revolving flats are replaced by sets of rollers called workers and strippers that are covered, normally with metallic saw-toothed cloth, but, on some older carding machines, with flexible wire, which interact with each other and the cylinder to open and separate the fibres. Woollen cards and carding machines used for the production of non-woven webs usually have two cylinders as well as associated workers and strippers.

In drafting, several card slivers are drafted or drawn on a drawframe (see Fig. 2.2) into one sliver to achieve better blending, and for blended yarns, slivers of different fibres can also be blended at this stage to meet the requirements of a more precise blending ratio. Drafting is performed at least twice — this helps to make the fibres more parallel and reduces the number of leading and trailing hooked fibres. Each drafting process is called a passage because the fibres pass through the drawframe. So a sliver that has been drafted twice would be referred to as a second passage drawframe sliver. The drafted slivers are then taken to the roving frame (see Fig. 2.3) or speed frame to produce the roving, which is an extremely long assembly of staple fibres, substantially parallel, slightly twisted, but still capable of being drafted in the later or final stages of preparation for spinning.

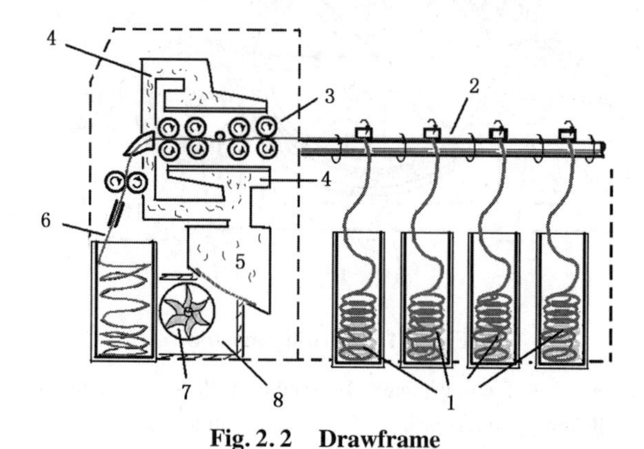

Fig. 2.2 Drawframe

1—card slivers 2—sliver guide roller 3—drafting assembly 4—suction dust catcher 5—dust filter box 6—drawn sliver 7—exhaust fan 8—air-exhausting pipe

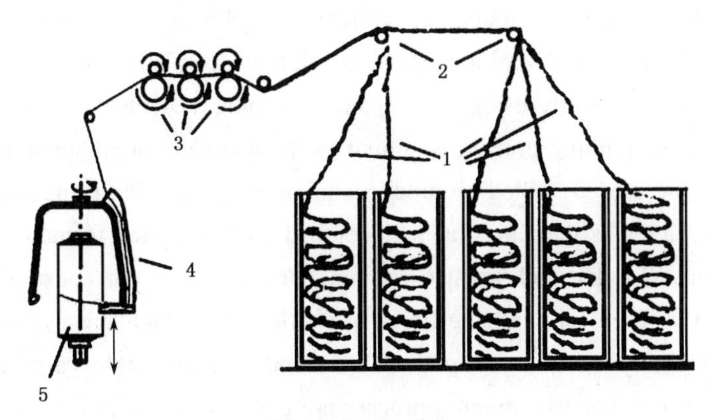

Fig. 2.3 Roving

1—drawn sliver 2—sliver guide roller 3—drafting assembly 4—flyer 5—roving bobbin

In the final spinning process, the roving is attenuated or drawn to the desired linear density (weight per unit length), which is the final drafting stage in the preparation for spinning, and the desired amount of twist is inserted, using a ring spinning system. In ring spinning the yarn package, mounted on a spindle, is rotated to insert twist into the fibres as they emerge from the front rollers of the drafting system on the machine. The twisted yarn is wound onto the package due to the drag imposed on the yarn which is threaded through a small metal traveller that slides around the ring as the spindle rotates (see Fig. 2.4).

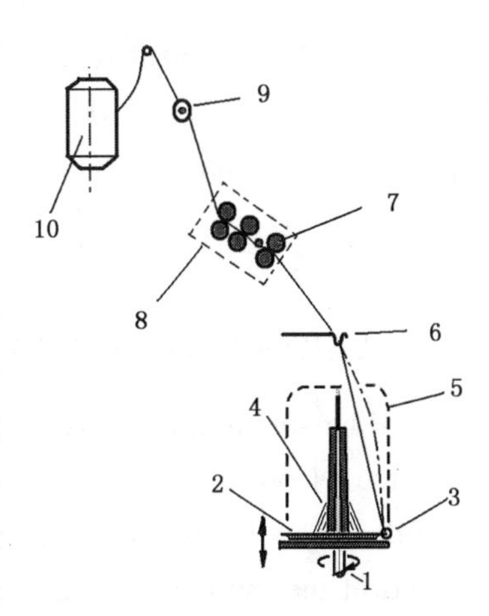

Fig. 2.4 Ring spinning

1—spindle 2—ring
3—traveller 4—bobbin
5—separating plate
6—guide 7—front rollers
8—drafting assembly
9—tensioner 10—roving

If a better quality yarn is desired, the card sliver, before drafting, can be put through a combing machine, which further straightens the fibres, and removes the short fibres and impurities so that finer yarns can be produced. Yarns produced with the additional combing process are called combed yarns, and yarns spun without being combed are called carded yarns. Combed yarns are used to produce lighter weight and better quality fabrics.

In woollen spinning, slubbings, which are lightweight continuous strands of staple fibres, instead of slivers are produced directly from the card. These are spun directly into yarns typically using ring or mule spinning systems.

The worsted spinning process is similar to the combed cotton spinning process in that the yarns are carded, gilled (a process similar to drafting but using pins to support and comb the fibres as they are drawn), combed, gilled again, formed into a roving and then spun. Woollen and worsted spinning processes are not constrained to wool fibres but can be used for synthetic fibres of similar length and fineness to wool fibres that would normally be spun using them. Normally, long staple yarns combed and produced using the worsted spinning process are called worsted yarns, and long staple yarns produced on the woollen system without combing are called woollen yarns. The worsted spinning process can be described as follows:

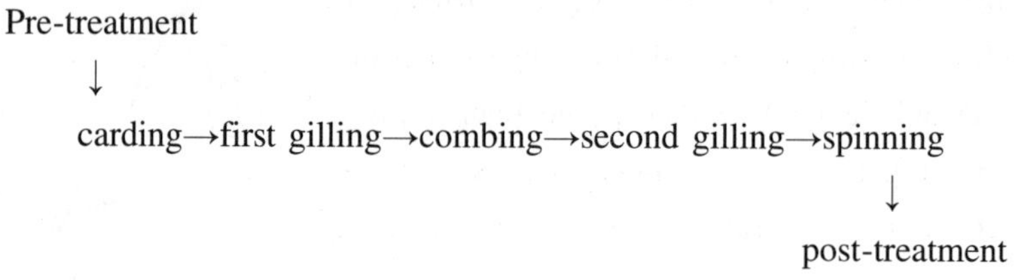

2.2 YARNS FROM OTHER SPINNING TECHNIQUES

Ring spinning has several limitations. The size of the ring may not be very large and the friction between the traveller and the ring limits the speed of rotation of the spindle. This limits both the size of the package that can be spun and the production speed of the machine and has greatly constrained the development of the ring-spinning system.

However since the ring-spinning mechanism is relatively simple, research has been undertaken to develop the ring-spinning, and the production of Sirospun and Sirofil yarns based on the ring-spinning techniques are two examples. In the former, two strands of roving from the front rollers enter into the same yarn guide and are twisted and wound on the same spindle to produce a two-ply yarn with the same direction of twist in both strands. In the latter, a filament is fed together with the roving from the front roller. Both are twisted and wound on the same spindle to produce a yarn with a filament core and with staple fibres wrapped around it.

Modern spinning technologies provide many new techniques for yarn spinning, and one commercially successful spinning technique is the open-end (OE) spinning. In the open-end spinning system, the roving process used in ring spun yarn production is eliminated. Typically, a second passage drawn sliver is fed into the open-end spinning machine, and the fibres are separated and joined to an "open end" of the yarn being produced. Twist for yarn strength is also inserted in this process. The commonly used types of open-end spinning are rotor spinning and friction spinning. In rotor spinning, yarn formation and twist insertion are achieved by depositing the fibres onto the inner wall of a rotating rotor and drawing the yarn from the wall through a nozzle mounted in the centre of the rotor (see Fig. 2.5). The positive rotation of the grooved winding drum rotates the yarn package, and the yarn guided in the groove moves to and fro along the axial direction of the yarn package and is wound onto it.

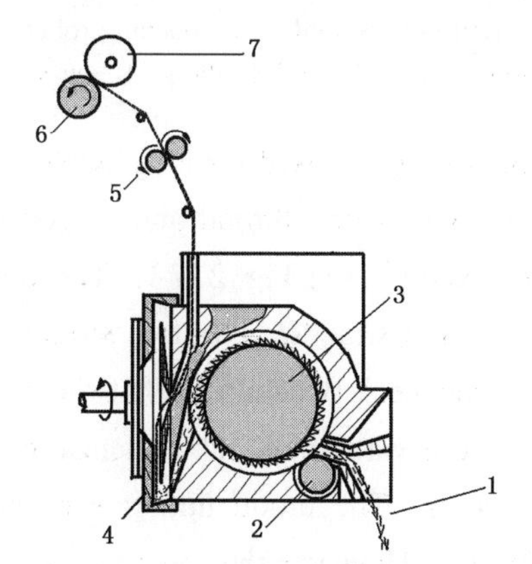

Fig. 2.5 Rotor spinning

1—drawn sliver 2—feeding roller 3—opening roller 4—rotor
5—outlet rollers 6—grooved drum 7—yarn package

In friction spinning, fibres are deposited between a pair of closely spaced friction drums, at least one of which is a suction drum with an arrangement of suction holes. Yarn is spun by pulling the fibres axially from the drums which insert twist into yarn through the frictional forces on the fibres as the drums rotate (see Fig. 2.6). In addition to rotor spinning and friction spinning, yarns have also been produced by air-jet spinning and vortex spinning. Open-end spinning system allows larger package sizes to be spun since twisting and winding are separated. Also, it is easier to rotate the small open-end of the yarn than it is to rotate the whole yarn package as it is done in ring spinning. This allows higher productivity and energy savings.

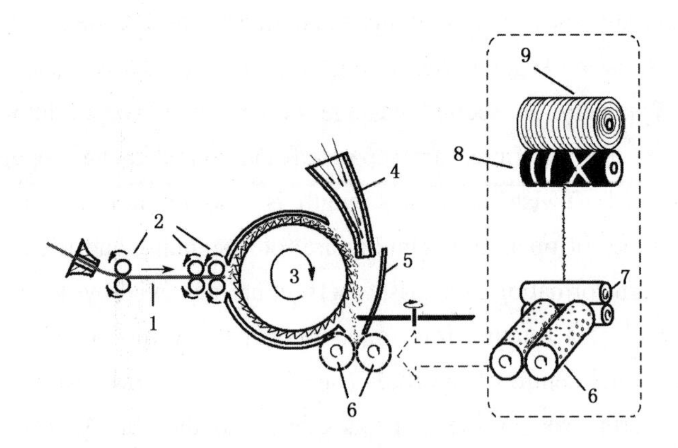

Fig. 2. 6 Friction spinning

1—drawn sliver 2—drafting assembly 3—opening roller 4—blow-in pipe 5—retaining plate 6—suction drums 7—outlet rollers 8—grooved drum 9—yarn packages

Another popular spinning system, used for long staple worsted type yarns, is self-twist spinning in which two rovings are drafted and passed through a pair of rotating rollers that also reciprocate axially (see Fig. 2. 7). The reciprocating motion inserts alternating S and Z twist into the strands. The two strands are brought together and twist around each other to produce a two-fold, self-twist (ST) yarn with alternating lengths of S and Z twist. These yarns may subsequently be twisted again to produce self-twist, twisted (STT) yarns. The production rate of this method of spinning is much higher than ring spinning. However the twist variation in the yarn can produce unwanted patterning effects in some fabrics.

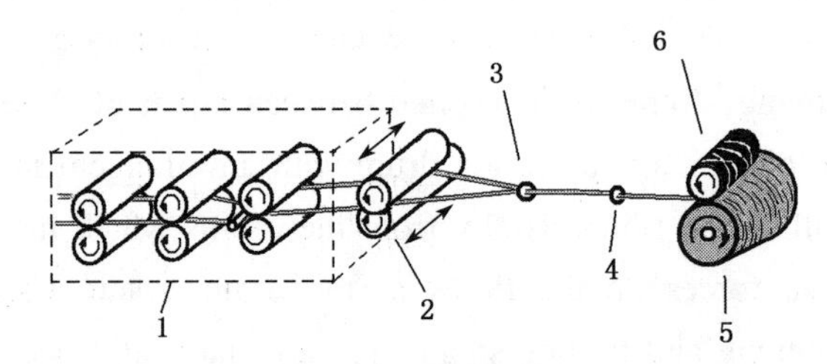

Fig. 2. 7 Self-twist

1—drafting assembly 2—reciprocating and rotating rollers
3—convergence guide 4—roving guide 5—yarn package 6—grooved drum

2.3 PLIED YARNS

Twisting two or more single yarns together will produce a plied or folded yarn or thread. Research has shown that the number of these single-ply (or singles as they are often referred to) yarns involved in the twisting process should be less than five, or the structure of the plied yarn will not be stable. Two-ply, or two-fold, yarns are commonly produced for weaving or knitting, but sewing threads generally use a three-ply structure. When a plied yarn is produced, the twist direction of the plied yarn generally differs from that of the component single-ply yarns. Typically S-twisted single yarns are twisted in a Z direction to form the plied yarn.

3 THE YARN QUALITY

Yarns can be classified in many ways, for example, as pure yarns or blended yarns according to whether two or more types of fibres were mixed during spinning; carded yarns or combed yarns according to whether a combing process was involved; singles yarns, ply-yarns, monofilaments, multifilaments, bulked yarns, textured yarns, wrap yarns, core yarns and slub yarns according to the morphology or yarn structure; ring-spun yarns, Sirospun yarns and open-end (in most cases rotor spun) spun yarns, etc., according to the spinning techniques used.

With developments in spinning technology and textile raw materials, there are more and more kinds of yarns appearing on the market.

Yarn quality is an important issue. The quality of yarns is evaluated from many aspects. These include yarn linear density and its deviation or irregularity, twist and its unevenness, yarn strength, specifically tensile strength and loop strength, and elongation at break, etc. Furthermore, visual yarn faults need to be observed and then evaluated.

3.1 YARN LINEAR DENSITY

The linear density of a yarn is the specification that corresponds to yarn thickness. There are two major systems used to calculate it: one is the weight based (fixed length) system and the other is the length based (fixed weight) system. The same methods can be used to define the fineness of the textile fibres.

3.1.1 Weight Based System

Tex is the standard unit in this system defined by the International Organization for Standardization (ISO) for the designation of yarn linear density. Tex equals the conditioned weight in grams of one thousand metres of yarn. For finer yarns, filaments or even fibres, the unit of decitex (dtex) is commonly used, which is one tenth of the tex.

Traditionally, denier, which is one ninth of the tex, is also used to define the linear density of filaments. That is to say, the denier of a filament is equal to the conditioned weight in grams for nine thousand metres of that filament.

It follows that, in the weight based system, the larger the value of the tex or denier, the higher its linear density will be and, for a given fibre density, the thicker the yarn will be.

3.1.2 Length Based System

Cotton count (or, English count) is a common unit in this system used in international trade for cotton yarns or cotton type yarns. The number of cotton count is defined as the number of 840 yard lengths of conditioned yarn that weigh one pound. Woollen and worsted counts are also defined on a similar basis but using different yardages of yarn. The former uses 256 yard lengths and the latter uses 560 yard lengths.

A similar unit to cotton count is the metric count, which is defined as the number of metres in length of a conditioned yarn that weigh one kilogram.

In the length based system, the smaller the count number is, the higher its linear density and the thicker the yarn will be.

If two plies of single yarns, say 14 tex, are twisted into a plied yarn, the linear density of the plied yarn is expressed as 14 × 2 tex, but the linear density of a ply-yarn formed by two plies of single yarn, say each with 42's cotton count, will be expressed as 42/2 count. For a multifilament, its linear density and the number of filaments in the yarn are defined. For example, a 75dtex/30F specifies a multifilament yarn composed of 30 single filaments with a resultant linear density of 75 dtex.

3.2 YARN TWIST

Twist is required to give the cohesion pressures among fibres within a yarn, so that when a tensile force is applied along the axis of the yarn, friction forces between the fibre surfaces are built up, which in turn reduce fibre slippage, giving the yarn a certain strength and extensibility. Twist can be made as left handed (S) direction, or right handed (Z) direction.

Twist level denotes the degree of twist, and it is defined as the number of twists or turns per unit length of a yarn. The more turns per unit length, the greater will be the cohesive forces between the fibres up to a critical point. Further increase in twist beyond a given value, that depends on both the fibre type and density and yarn construction and density, would increase the helix angle (or twist angle) of the fibres within the yarn, such that the breaking strength of each fibre can not be efficiently realised, which could in fact make the yarn weaker.

The twist level is measured as the number of twists applied in a certain yarn length, say one metre, from which the twist factor, which is deduced from the said twist level and the yarn linear density, may be defined. Both the twist level and the twist direction are important, which need to be specified when ordering yarn. For example a specification that defines a singles yarn with linear density of 40 tex and 640 twists in one metre in the S direction is 40 tex S640.

3.3 YARN STRENGTH AND ELONGATION

Yarn strength, especially the tensile strength and associated breaking elongation, are the important mechanical parameters that need to be controlled. Yarn breakage is caused by fibre slippage and fibre breaking, and therefore the characteristics of the fibre, the linear density and the twist of the yarn affect the yarn strength. Besides, the greater the irregularity in the linear density or twist is, the weaker the yarn would be, since the yarn breaks at its weakest point. Furthermore, variation in yarn linear density may manifest itself in fabrics, as, for example, a stripy appearance, particularly if the variation is periodic and therefore affects the appearance of the fabric as well as its mechanical properties.

Yarns are spun for a specific application and hence the quality requirements will differ accordingly. For example, hosiery yarns (yarns for knitting) should be softer and more flexible than yarns used in weaving, and so their twist level tends to be lower. Furthermore the twist direction in hosiery yarns used in circular knitting needs to be carefully selected.

For functional yarns, some particular properties must be further observed. One example is that electrical properties such as electrical resistance would have to be measured and evaluated for the electro-conductive yarns to be used for smart textiles.

When purchasing yarns, the buyer should clearly specify the precise requirements, defining the type of the yarn and the detailed specifications with allowable tolerances. A specification should include: the type of fibre, blend ratio (if blended yarns are requested), the linear density, twist (or twist factor) and twist direction. Furthermore, the spinning technique to be used, the breaking strength, limitation on irregularity, etc. would also need to be specified.

The supplier should evaluate whether the buyer's specification is workable and the tolerance of the yarn specifications should be clarified with the buyer if necessary, since it is impossible to produce a yarn with an exact and absolute value of twist or breaking strength, etc.

Words and Phrases

yarn	纱
knitted fabric	针织物
woven fabric	机织物
staple fibre	短纤
continuous filament	长丝
monofilament [ˈmɒnəʊˈfɪləmənt]	单丝
multifilament [ˌmʌltɪˈfɪləmənt]	复丝
spin(过去分词 spun)	纺丝,纺纱
spinning	纺纱或细纱(工艺),纺丝(工艺)
melt spinning	熔体纺丝
solvent spinning	溶液纺丝
decompose [ˌdiːkəmˈpəʊz]	分解
polymer chip	聚合物切片
spinneret [ˈspɪnəret]	喷丝头
solidify [səˈlɪdɪfaɪ]	固化

melting temperature 熔点
carbonise [ˈkɑːbənaɪz] 炭化
dry spinning 干法纺丝
wet spinning 湿法纺丝
molecular chain 分子链
bundling [ˈbʌndlɪn] 集束
washing 水洗
oiling 上油
crimping 卷曲
heat setting 热定形
delustrant 消光剂
titanium [taɪˈteɪnjəm, tɪ-] dioxide[daɪˈɒksaɪd] 二氧化钛
dull [dʌl] fibre 消光纤维
semi-dull fibre 半消光纤维
bicomponent [ˌbaɪkəmˈpəʊnənt] fibre 双组分纤维
micro fibre 超细纤维
unconventionally shaped fibre 异形纤维
twisting 加捻
winding [ˈwaɪndɪŋ] 络纱
textured 经变形的
false-twisting 假捻
de-twisting 退捻
air-texturing technique 空气变形技术
spun yarn 短纤纱
blended yarn 混纺纱
yarn spinning 纺纱
ring spinning 环锭纺
spinning mill 纺纱厂
in bales 以包的形式
wire-toothed cloth 钢丝针布
carding machine 梳棉机
sliver [ˈslɪvə, ˈslaɪvə] 棉条,纱条

card sliver	粗梳生条
lap	棉卷,纤维卷
opening	开棉
cleaning	清棉
mixing [ˈmɪksɪŋ]	混棉
blending [blendɪŋ]	混棉
lap rod	棉卷扦,纤维卷扦
lap holder	棉卷架,纤维卷架
lap drum	棉卷罗拉
unwind [ˈʌnˈwaɪnd]	退绕
feeding plate	给棉板
feeding roller	给棉罗拉
licker-in [ˈlɪkəɪn]	刺辊
taker-in	刺辊
tufts of fibre	纤维簇
centrifugal [senˈtrɪfjʊgəl] force	离心力
mote [məʊt] knife	除尘刀
licker-in under-casing	刺辊下壳罩(小漏底)
flat	盖板
carding	粗梳
cylinder [ˈsɪlɪndə]	锡林(粗梳机大滚筒)
cylinder under-casing	锡林下壳罩(大漏底)
flat strip [strɪp]	盖板花
flat stripping comb	上斩刀
flat stripping brush	盖板刷帚
doffer [ˈdɒfə]	道夫(粗梳机小滚筒)
stripping roller	剥棉罗拉
card	粗梳,粗梳机
nip roller	轧辊
trumpet [ˈtrʌmpɪt]	喇叭口
chute [ʃuːt]	(斜向)管道
woollen and worsted cards	粗纺和精纺毛纱的粗梳机

workers	梳毛辊
strippers	剥离辊
can coiler [ˈkɔɪlə]	圈条器
sliver can	生条筒
drafting	牵伸
drawframe [ˈdrɔːfreɪm]	并条机
blending ratio	混纺比
leading hooked fibre	前弯钩纤维
trailing hooked fibre	后弯钩纤维
second passage drawframe sliver	二道熟条
sliver guide roller	导条辊
drafting assembly [əˈsemblɪ]	牵伸装置
suction dust catcher [ˈkætʃə]	吸尘箱
dust filter box	滤尘箱
exhaust [ɪgˈzɔːst] fan	排风扇
air-exhausting pipe	排风管
drawn sliver	熟条
roving [ˈrəʊvɪŋ]	粗纱
roving frame	粗纱机
speed frame	粗纱机
flyer [ˈflaɪə]	锭翼
roving bobbin [ˈbɒbɪn]	粗纱筒子
linear [ˈlɪnɪə] density	线密度
twist	加捻
yarn package	纱的卷装
spindle [ˈspɪndl]	锭子,芯轴
front roller	前罗拉
traveller	钢圈
ring	钢领
separating plate	隔纱板
combing [ˈkəʊmɪŋ] machine	精梳机
combing process	精梳加工

combed yarns	精梳纱
carded yarn	普梳纱
slubbing [ˈslʌbɪŋ]	头道粗纱
mule [mjuːl] spinning	走锭纺纱
worsted [ˈwʊstɪd] spinning	毛(型)纱精纺
gilling [ˈgɪlɪŋ]	针梳
woollen [ˈwʊlɪn] spinning	毛(型)纱粗纺
worsted yarn	精纺毛(型)纱
woollen yarn	粗纺毛(型)纱
Sirospun yarn	赛络纱
Sirofil yarn	赛络菲尔纱
open-end (OE) spinning	自由端纺纱
rotor [ˈrəʊtə] spinning	转杯纺纱
friction spinning	摩擦纺纱
nozzle [ˈnɒzl]	喷嘴
grooved [gruːvd] winding [ˈwaɪndɪŋ] drum	络纱槽筒
grooved drum	(络纱)槽筒
feeding roller	喂给罗拉
opening roller	分梳辊
rotor	转杯
outlet [ˈaʊtlet, -lɪt] rollers	输出罗拉
suction drum	吸风滚筒
blow-in pipe	吹风管
retaining [rɪˈteɪnɪŋ] plate	挡板
air-jet spinning	喷气纺纱
vortex [ˈvɔːteks] spinning	涡流纺纱
self-twist spinning	自捻纺纱
S twist	S 捻
Z twist	Z 捻
self-twist (ST) yarn	自捻纱
self-twist, twisted (STT) yarns	加捻自捻纱
reciprocating [rɪˈsɪprəkeɪtɪŋ] and rotating rollers	搓捻辊

convergence [kən'vɜːdʒəns] guide	汇合导纱钩
plied yarn	股纱
folded yarn	股纱
thread	线
singles	单股纱
singles yarn	单股纱
ply-yarn	股纱
bulk yarn	膨体纱
textured yarn	变形纱
wrap [ræp] yarn	包覆纱
core [kɔː] yarn	包芯纱
slub [slʌb] yarn	竹节纱
raw [rɔː] material	原料
linear density deviation [ˌdiːvɪ'eɪʃən]	线密度偏差
twist unevenness ['ʌn'iːvənɪs]	捻度不匀
weight based system	定长制
length based system	定重制
International Organization for Standardization (ISO)	国际标准化组织
tex[teks]	特(克斯)
conditioned weight	公定回潮率重
decitex [desɪˌteks] (dtex)	分特(克斯)
denier ['denjə(r)]	旦尼尔
cotton count	英制棉纱支数
English count	英支
woollen count	粗纺毛(型)纱支数
worsted count	精纺毛(型)纱支数
metric count	公支
resultant [rɪ'zʌltənt] linear density	总线密度
cohesion [kəʊ'hiːʒən] pressure	抱合力
fibre slippage	纤维滑移
twist level	捻度
cohesive [kəʊ'hiːsɪv] force	抱合力

twist factor	捻系数
periodic [pɪərɪˈɒdɪk]	周期性的
hosiery [həʊzɪərɪ, ˈhəʊʒərɪ] yarns	针织用纱
functional yarn	功能纱线
electrical resistance	电阻
electro-conductive yarn	导电纱线
smart textiles	智能纺织物

Exercises

1. Continuous polyamide filaments are spun through ________. (　　)
   a) dry spinning　　b) melt spinning
   c) ring spinning　　d) wet spinning
2. The main purpose of the drawing process in manufacturing polyester filaments is to make them ________. (　　)
   a) longer　　b) finer
   c) smoother　　d) stronger
3. The reason why polyester is made by melt spinning instead of solvent spinning is that ________. (　　)
   a) it cannot be dissolved in any solvent
   b) solvent spinning usually causes environment pollution
   c) it has a clearly defined melting temperature
   d) no suitable liquor can solidify the extruded filament
4. Which of the following processes are not used in the production of carded yarns? (　　)
   a) Drafting　　b) Carding　　c) Roving　　d) Combing
5. Which of the processes given below will be involved in the manufacture of worsted yarns? (　　)
   a) Roving　　b) Gilling　　c) Combing　　d) Drafting
6. In which of the following spinning systems is the twist in the yarn inserted through the reciprocating movement of a pair of rotating rollers? (　　)

a) Ring b) Rotor c) Self-twist d) Friction

7. Which of the following spinning systems does not involve roving production? ( )

a) Ring b) Rotor c) Air-jet d) Friction

8. Which of the following systems used to define the fineness of textile fibres is categorised as a weight based system? ( )

a) Tex b) Cotton count c) Denier d) Metric count

9. A polyester multifilament yarn with a resultant linear density of 150 dtex composed of 32 single filaments can be defined as ________. ( )

a) 150/32dtex b) 150dtex/32F

c) 32F/150dtex d) 150dtex(32)

10. The common way to specify a singles yarn with linear density of 40 tex and with 640 twists in one metre in the S direction is ________. ( )

a) 40 (640) tex b) 40/ 640 tex

c) 640S/40tex d) 40 tex S640

Reading Materials

Yarns Made from Bamboo Viscose, Greater Bariety in Sustainable Materials

COVID-19 is confining people all over the world to their homes and causing demand for comfortable leisurewear to soar. As a result, product developers are working hard as well. In order to cater to the demand for feel-good outfits, a cosy terry fabric has been created using a yarn obtained from bamboo cellulose, thus also reflecting the trend towards sustainability. The natural fibre material forms soft, absorbent loops that are densely arranged on the inside of the textile. The outside is made of a polyester microfibre yarn that creates an ultra-soft fleece layer.

A sporty set comprising a hooded jacket and jogging trousers — made from this innovative double-faced material — was presented at KARL MAYER's booth at ITMA ASIA + CITME in June 2021, where it attracted a great deal of attention. The warp knitted terry cloth was also an important topic at a company webinar held shortly after the exhibition.

— excerpted from *International Textile Market*, November-December 2021, page 26

【参考提示】

1. *International Textile Market* 是印度 Times International 公司出版的双月刊杂志。从文中 cosy, fibre 等可以看到本文作者习惯英式拼法。

2. terry fabric,毛圈布,因此下文提及的"soft, absorbent loops"中的 loops 指毛圈布上的毛圈。

3. ultra-soft fleece layer,超软的起绒层。所提及的织物,一面是毛圈,另一面是绒面,所以下文称之为 double-faced material。关于经编的基本知识,可参看课文第 4 章。

4. Karl Mayer, 国内常译为"卡尔迈耶",经编机制造上非常出名的德国公司。

5. booth, 展位。

6. ITMA ASIA + CITME,中国国际纺织机械展览会暨国际纺织机械展览会亚洲展会。ITMA 源于德文 Internationale Textilmaschinen Ausstellung,Ausstellung 即"展览会"。ITMA 是四年一度的国际纺机博览盛会。本文提及的是 ITMA 旗下的展览会。

7. Webinar, 线上研讨会。

Eco-Conscious Cellulosic Fiber Yarn Realized with Rieter Expertise

Eastman Chemical engaged Rieter when it wanted to extend its Naia™ cellulosic product into staple fiber yarns

TW Special Report

Global specialty chemicals company Eastman Chemical Co. , Kingsport, Tenn. , turned to Switzerland- based Rieter regarding a process consultancy for its new staple fiber. The cellulosic fiber Naia™ already existed as a filament yarn and can now be utilized for a wider range of textile applications. Eastman and Rieter worked together to find the right blends and yarn counts for the Naia staple fiber to increase its market penetration, while giving fashion brands a truly sustainable choice.

— excerpted from *Textile World*, January/February 2022, page 30

【参考提示】

1. Eastman Chemical,国内有译为"伊士曼化工"的,美国知名化工材料公司。

2. Rieter,指瑞士的 Rieter Group, 国内有译为“立达集团”的,瑞士知名的短纤纱纺机企业。

3. *TW* Special Report,《纺织世界》的特别报道。

4. right blends and yarn counts, 正确的混棉及正确的纱支数。

Optimizing the Cotton Yarn Spinning Process

With the state-of-the-art comber machine TCO 21, Trützschler has expanded its position as a full-range supplier in spinning preparation. The TCO 21 meets the requirements of cotton yarn producers who are searching for more effective spinning processes and solutions that facilitate higher productivity, better yarn qualities and automatic optimization.

The big advantage of the TCO 21 is in its processing speed and its range of modern, easy-to-use automated features. The technology is designed to empower operators to customize and optimize performance quickly and is simple to handle even for employees with less experience — an important factor in times of staff shortages and high staff costs.

— excerpted from *International Fiber Journal*, Issue 1 2022, page 10

【参考提示】

1. Trützschler,德国知名的纺机制造商,国内有译为“特吕茨施乐”的。TCO 21 是它提供的精梳机型号。

2. state-of-the-art,最先进的。

CHAPTER 3

WEAVING AND WOVEN FABRICS

1 FUNDAMENTAL FEATURES OF A WEAVING MACHINE

It is believed that the weaving machine, or loom, has been in use since 4400 BC. For many years, weaving technology and weaving machinery have seen tremendous developments. However, irrespective of whether they are modern or traditional, all looms have the same fundamental features. These are the let-off device, the shedding device, the weft insertion device, the beating-up device, and the fabric take-up and batching device (Fig. 3. 1).

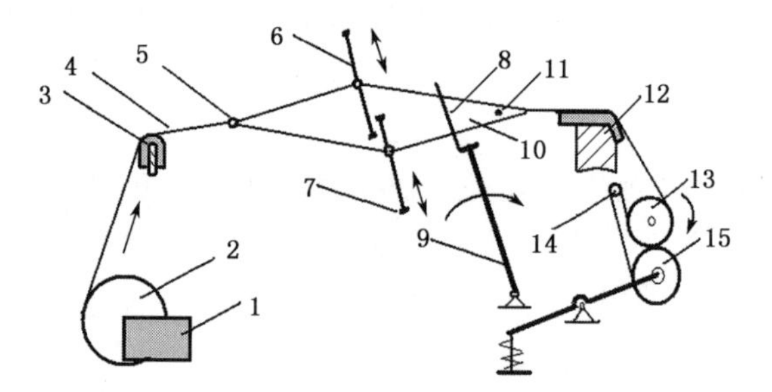

Fig. 3. 1 Illustration of primary weaving

1—let-off device 2—loom beam 3—back rest 4—warp 5—lease rod
6—heald 7—heald shaft 8—reed 9—sley 10—shed 11—weft
12—breast beam 13—take-up roller 14—guide roller 15—batching roller

The let-off device is the device that gradually releases the warps on the beam containing them to move towards the weaving position. A traditional loom uses the negative let-off system, whereby the warps are pulled from the beam by the forces from the weaving area which are mainly the forces due to the fabric take-up mechanism. On modern machines, a positive let-off device is used, whereby the speed of the warp let-off run-in is positively controlled and varied by mechanical or electronic means. This is achieved through the use of a sensor on the breast beam to detect the warp tension, and through a control circuit, the signals from the sensor adjust the warp let-off and cloth take-up motors.

The shedding device is the device, which separates or opens groups of warp yarns (or sheets as they are often called) to form a passageway or shed to let the weft pass through. On a loom, there are heald shafts (referred to harnesses in the USA), into which the healds (or, heddles in the USA) are set evenly. Each heald has an eye through which each warp yarn is threaded. As a heald shaft moves up, the warps passing through its healds will raise. Conversely, as the heald shaft moves down, the warps will lower. The simplest loom uses two heald shafts, which move up and down alternately to form a shed. It can be seen that the more heald shafts a loom has, the more complex are the structures that the loom can produce. A typical, simple weaving loom has 2 to 8 heald shafts controlled by cams or tappets, and, if a loom has a dobby to control the movement of the heald shafts, it could have 16 to 32 heald shafts programmed either by a punched paper chain or, more commonly nowadays, an electronic control system, to produce fabrics with more complicated structures. If a jacquard mechanism is used, the raising and lowering of every warp can be controlled individually without using heald shafts, and very intricate designs with very large pattern repeat could be made.

The weft (or filling) insertion device inserts the weft yarn through the shed. On an older style, traditional loom, a shuttle is used. When the shed is formed, a picking stick hits the shuttle in the shuttle box. Prior to shuttle weaving the weft yarn is wound onto a pirn. The shuttle carries the pirn with the weft, passing through the shed, and is caught in the shuttle box at the other side of the loom to insert the weft into the fell of the fabric. The slow back and forth reciprocating movement of the large mass of wooden shuttle through the shed and its limited yarn carrying capacity greatly hinder the productivity and it is difficult to control the weft tension. Furthermore, the picking and catching action of the shuttle makes the loom noisy. Therefore, in most modern mills, shuttleless looms, such as rapier looms, air or water jet looms or projectile looms, are used. The great advantage of these machines is that, unlike the shuttle loom, it is not necessary for the yarn packages to be transported though the shed to insert the weft.

On the rapier loom, depending on its width, either a single rapier or two rapiers (an entry rapier and an exit rapier) are used (Fig. 3.2). These may be flexible or rigid. In the more common double rapier looms, when the shed is formed, both rapiers enter the shed, with the entry rapier holding the weft yarn and pulling it from a stationary yarn package at the side of the loom. When they meet in the middle across the cloth, the exit rapier collects the weft yarn from the entry rapier, and then, both rapiers withdraw. The withdrawal of the exit rapier completes the weft insertion.

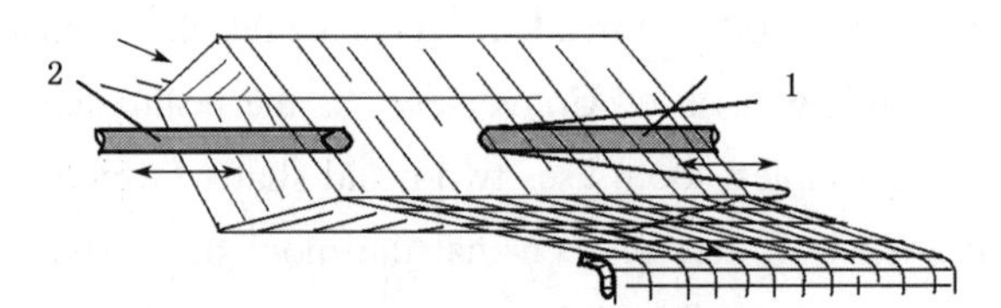

Fig. 3.2 Rapier weft-insertion

1—entry rapier 2—exit rapier

On the air-jet loom, the weft yarn is carried in a jet of air through the shed (Fig. 3.3). Weft yarn from the stationary yarn package passes through a yarn tensioner and enters into a weft measuring and storing device, which measures the required length of weft and keeps it in a relaxed state. Then the weft yarn is introduced into the nozzle. A clamp situated between the measuring and storing device and the nozzle holds the weft. When the shed is formed, the clamp releases the weft and the high pressure air-flow from the nozzle pulls the weft from the storage device out and drags it through the shed. After the weft has been inserted, the clamp holds the weft again, and a clip before the nozzle cuts the weft in readiness for the next insertion. On some machines,

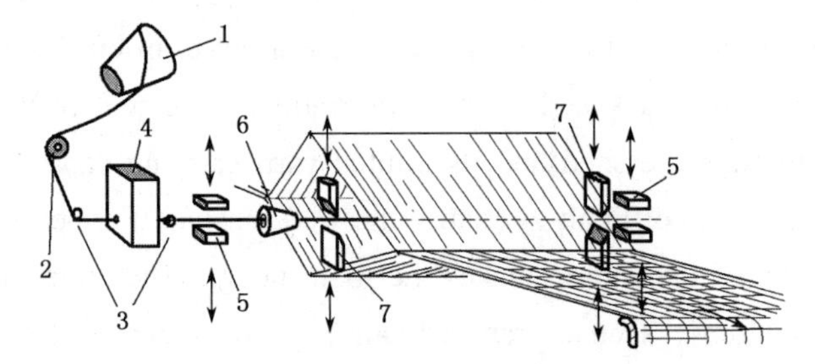

Fig. 3.3 Air-jet weft-insertion

1—yarn package 2—tensioner 3—yarn guider
4—yarn measuring and storing device 5—clamp 6—nozzle 7—scissors (clip)

auxiliary nozzles or relay nozzles are positioned across the loom to help the weft yarn to cross the shed. Some jet machines make use of jets of water to force the weft through the shed.

On a projectile loom a projectile is used (Fig. 3. 4). Like the two types of loom previously described, the projectile carries only a single length of weft rather than a pirn through the shed. The projectile carrying the weft is shot across the shed and then conveyed back, beneath the fabric, for a subsequent insertion.

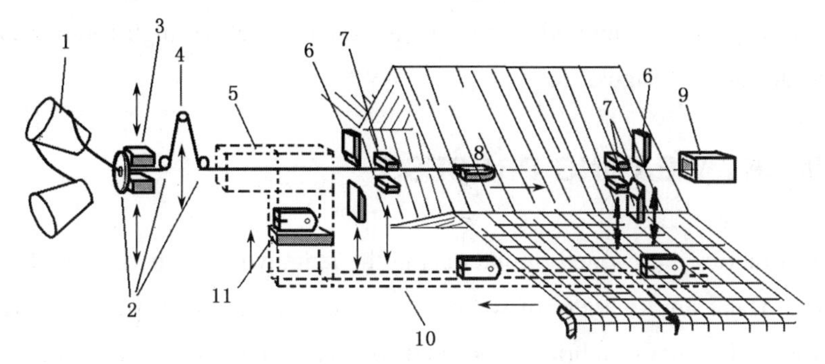

Fig. 3. 4 Projectile weft-insertion

1—yarn packages 2—yarn guide 3—weft clamp 4—tension lever
5—weft feeder and projectile feeder 6—weft end scissors 7—weft end clamp
8—projectile 9—projectile catching-brake 10—projectile conveyer 11—projectile lift

The shuttleless looms have a much higher production rate than a shuttle loom. However, because the weft is always inserted from the same side of the loom, either a fringed selvedge is produced, which needs to be cut off, or selvedge tuck-in devices need to be installed to tuck the ends of the weft back into the cloth after weft insertion (Fig. 3. 5).

The beating-up device beats the inserted weft into the fell of the cloth. The device comprises a reed, which is a comblike frame, mounted on a sley. All warps, often in groups, pass through the evenly spaced gaps, called dents, between the reed wires. As each weft yarn is inserted, the sley swings forward, and the reed pushes the weft yarn firmly against the fell of the fabric to produce the cloth.

Fig. 3. 5 Tuck-in selvedge

The fabric take-up and batching device

comprises a series of rollers. The positive rotation of the take-up roller pulls the cloth from the weaving area and winds it onto the batching roller. The take-up force is also the main factor to keep the warp taut. The tensions in the warp and the fabric, the let-off and take-up rates, together with the linear density of the weft yarns, determine the pick (or filling) density within the fabric.

On a modern weaving machine, some of the above mentioned devices are electronically controlled. For example, with an electronic shedding device, the dwell and cross timing of each individual frame can be easily set for high running efficiency with intricate fabric designs.

2 PRIMARY WEAVING PROCESS

Woven fabrics are formed by interlacing the lengthwise warp yarns and widthwise weft yarns. There are a variety of woven structures, but as far as the weaving process is concerned, a woven fabric is formed by the following steps: warp let-off, shedding, weft insertion, beating-up and fabric taking-up.

To produce a plain woven fabric, which is the simplest woven structure, warps from the loom beam pass over the back rest and around lease rods and then threaded alternately through the healds on two heald shafts, to form two warp sheets. The let-off device releases the warp sheets under tension and the shedding device raises or lowers each heald shaft respectively to form the shed. The weft yarn, or filling yarn, can now be inserted across the shed by the weft insertion device, and then the reed, mounted on the sley, beats the weft against the cloth fell to compress it, and the inserted weft now becomes part of the cloth. The take-up roller rotates continuously, pulling the newly formed cloth away from the weaving position, and then the cloth is wound onto the batching roller (see also Fig. 3.1).

3 PREPARATORY WEAVING PROCESSES

Before weaving, it is necessary to wind the warp yarns onto the warp beam, a process referred to as warping. Hundreds of warp yarns are first wound from cones on stand creels onto a large warp beam by a warping machine. During this process the warp yarns are coated with a sizing compound (e. g. a starch mixture) to add strength and reduce yarn hairiness for weaving. The sized and dried yarns are then wound parallel to each other either as a single sheet or in sections across the width of the warp beam that will be ultimately be placed on the loom. For weft yarns, winding is necessary to prepare suitable packages for weaving. For shuttleless weaving, the yarns are wound onto cones. For shuttle weaving the yarns are wound onto pirns.

4 COMMON WOVEN FABRICS

There are many varieties of woven structures. Technologically, they can be classified as three types: primary weaves, derivative weaves and combination weaves.

4.1 PRIMARY STRUCTURES

There are three primary weave structures, which are plain weave, twill and satin (or sateen) (see Fig. 3.6).

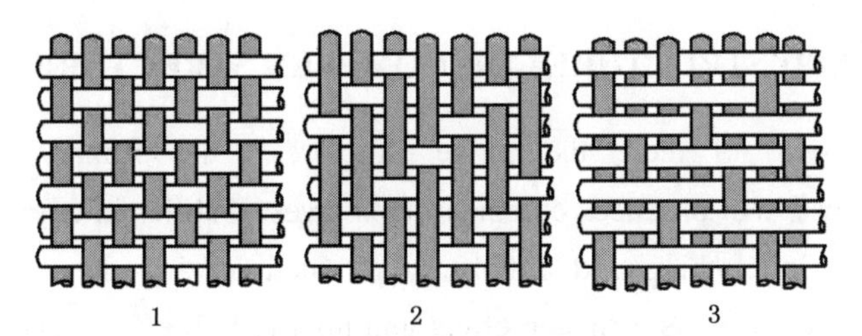

Fig. 3.6 Primary structures

1—plain weave 2—twill 3—(filling face) satin

Plain weave, the most common woven structure, is produced by making the warp and weft interlace under and over alternate weft and warp yarns respectively. Fabrics in this structure are widely used for cotton and polyester - cotton sheeting (bed linen) and shirtings. Poplin, chintz, valitin, palace, pongee, habotai and taffeta are commonly used examples of these. Pongee and taffeta used to be terms referring only to silk fabrics but nowadays nylon taffeta and polyester pongee are very common in the market.

Twill weave is produced by interlacing yarns to form diagonal lines or ridges across the fabric. This structure is used for sturdy products such as denim, gabardine, serge and khaki. They are commonly used as the shell fabrics for uniforms.

Satin can be divided into warp face satin and weft face (filling face) satin, which is called sateen. The most common is the warp face satin, in which the weft yarns cross over one and under several warp yarns, and thus mainly the warp yarns are visible on the face of the fabric. In a filling face satin, the filling yarns cross under one and over several warp yarns, and thus mainly the weft yarns are visible on the face. The satin weave structure produces a fabric with a smooth and lustrous surface. It is commonly used for upholstery, home decorating and fashionable apparel. Satin fabrics are commonly made with silk or man-made filament yarns.

4.2 DERIVATIVE STRUCTURES

As the name implies, derivative weaves are derived through some variations to the primary weaves. There are three types of derivative weaves. These are plain weave derivatives such as basket structures and oxford weaves, etc., twill weave derivatives such as diamond weaves and herringbones, etc., and the satin derivative weaves.

4.3 COMBINED STRUCTURES OR COMPLEX STRUCTURES

Combining two or more the above structures (the primary weaves or the derivative weaves) will produces combination weaves. Fabrics in combination weaves are widely used in clothing and upholstery production, and they generally have a special appearance, such as a ripstop effect and honeycomb effect, etc. If the warp or weft yarns are from two or more yarn systems, complex structures will be produced. Commonly used complex weaves are backed weaves (warp backed weave or weft backed weave), leno weaves, terry weaves and corduroy, etc. Fabrics in these complex structures are widely used for winter garments, upholstery cloths and industrial textiles.

5 QUALITY ISSUES OF THE WOVEN FABRICS

The cloth leaving the weaving machine and before being further treated is called the grey or greige cloth. Any weaving faults that appear in the grey cloth would affect the quality, especially the perceived or visual quality, of the finished fabrics.

Many visual weaving faults are due to uneven tensions on warps or wefts, which mainly result from an imperfect machine setting. For example, malfunctioning of the weft insertion device may cause some wefts to be inserted tightly and some slackly, so that weft "cracking" would be evident. If there is a variation in tension in the warp threads, warp "streakiness" may be observed. If the overall warp tension was set too high, warp "smashing" (where the warp breaks regularly during weaving) may happen. Uneven tension could also be caused due to insufficient control when preparing the weft yarn packages, or due to poor tension-setting during warping.

Many weaving faults are due to problems with machine components. Burrs (small rough pieces of protruding metal) on the healds or reed wires could mark the warp yarn, and heald marks or reed marks could occur. Reed marks might also arise from inappropriate denting.

Yarn faults in warp and/or weft would also affect the fabric quality. Highly irregular yarns or yarns containing slubs would affect both weaving efficiency and fabric appearance. Poor yarn tensile strength would lead to poor fabric strength. Attention should also be paid to the fact that mixing yarns (especially synthetic yarns or their blends) from different production lots might not show any apparent problems in the grey cloth but may present as faults in the fabric when the fabric is dyed and finished or cause problems during the processes themselves.

Of course, any incorrect setting up of the machines either before or during the weaving process will also cause problems. Misdrawing of the yarns through the healds, or, mistakes in setting weaving parameters or programming the looms would manifest as faults, so samples need to be checked carefully before full production is authorized. Furthermore, conditions in the weaving room, such as a dusty environment, high or low humidity, may cause loom fly. Excessive lubrication on the machine could lead to oil stains appearing on the cloth.

Words and Phrases

weaving machine 织机

loom [luːm] 织机

weaving technology	织造工艺
let-off device	送经装置
shedding [ˈʃedɪŋ] device	开口装置
weft insertion device	引纬装置
beating-up device	打纬装置
fabric take-up and batching device	织物牵拉卷取装置
heald[hiːld]	综丝,综片
heddle [ˈhedl] (*Ame.*)	综丝
weft	纬纱
loom beam	织轴
heald shaft [ʃɑːft]	综框
harness (*Ame.*)	综框
breast beam	胸梁
warp	经纱
reed [riːd]	钢筘
take-up roller	牵拉辊
back rest	后梁
sley [sleɪ]	筘座
guide roller	导辊
lease [liːs] rod	分经棒
shed [ʃed]	开口,梭口
batching [ˈbætʃɪŋ] roller	卷布辊
weaving position	编织位置
negative let-off system	消极送经系统
sensor	传感器
control circuit	控制电路
signal	信号
warp sheet	经纱片
dobby [ˈdɒbɪ]	多臂装置
jacquard [dʒəˈkɑːd] mechanism	贾卡提花机构
pattern repeat	花型循环
filling	纬纱

shuttle	梭子
picking stick	打梭杆
shuttle box	梭箱
shuttle weaving	梭织
pirn [pɜːn]	纡子
fell	织口
reciprocating movement	往复运动
hinder	阻碍
shuttleless loom	无梭织机
rapier [ˈreɪpɪə] loom	剑杆织机
air-jet loom	喷气织机
water-jet loom	喷水织机
projectile [prəˈdʒektaɪl; (*US*)-tl] loom	片梭织机
entry rapier	送纬剑杆
exit rapier	接纬剑杆
stationary[ˈsteɪʃ(ə)nərɪ]	固定的
tensioner	张力器
weft measuring and storing device	纬纱测长储纱装置
clamp	夹
clip	剪
auxiliary nozzle	辅助喷嘴
relay [ˈriːleɪ] nozzle	辅助喷嘴
projectile	片梭
yarn guide	导纱器
weft clamp	夹纬装置
tension lever	张力杆
weft feeder	喂纬器
projectile feeder	喂梭器
weft end scissors	纬纱纱尾剪
weft end clamp	纬纱纱尾夹
projectile catching-brake	接梭制梭器
projectile conveyer	传梭器

projectile lift	提梭器
fringed [frɪndʒd] selvedge ['selvɪdʒ]	毛布边
selvedge tuck-in device	布边折入装置
tuck-in selvedge	折入的布边
comblike	梳状的
dent [dent]	筘隙
reed wire	筘片
pick density	纬密
filling density	纬密
dwell and cross timing	停顿和交错的时间设定
frame	综框
interlacing	交织
plain woven fabric	平纹织物
warping	整经
cone	圆锥型纱筒
stand creel	落地纱架
sizing compound	上浆复合浆料
starch mixture	淀粉混合物
sized and dried yarns	经上浆并烘干的纱
warp beam	经轴
primary weaves	基本机织组织
derivative [dɪ'rɪvətɪv] weaves	变化机织组织
combination weaves	复合机织组织
plain weave	平纹
twill	斜纹
satin ['sætɪn]	缎纹,经面缎纹
sateen [sæ'tiːn]	纬面缎纹
cotton sheeting	棉粗平布
polyester-cotton sheeting	涤/棉粗平布
bed linen	床单
shirting	细平布,衬衣料
poplin ['pɒplɪn]	府绸

chintz [tʃɪnts] 有光布,轧光布
Valitin 凡立丁
Palace 派力司
pongee [pɒnˈdʒiː] 茧绸
Habotai 电力纺
taffeta [ˈtæfɪtə] 塔夫绸
nylon taffeta 尼丝纺
polyester pongee 春亚纺
denim [ˈdenɪm] 牛仔布
gabardine [ˈgæbədiːn] 华达呢
serge [sɜːdʒ] 哔叽
khaki [ˈkɑːkɪ] 卡其
shell fabric 面料
warp face satin 经面缎纹
weft face satin 纬面缎纹
filling face satin 纬面缎纹
upholstery [ʌpˈhəʊlstərɪ] 室内装潢
fashionable [ˈfæʃənəbl] apparel [əˈpærəl] 时装
plain weave derivatives 平纹变化组织
basket structures 方平组织
oxford [ˈɒksfəd] weave 重平组织
twill weave derivative 斜纹变化组织
diamond weave 菱形斜纹
herringbone [ˈherɪŋbəʊn] 人字斜纹
satin derivative weave 缎纹变化组织
ripstop effect 方格效应
honeycomb [ˈhʌnɪkəʊm] effect 蜂巢效应
backed weaves 二重组织
warp backed weave 经二重组织
weft backed weave 纬二重组织
leno [ˈliːnəʊ] weave 机织纱罗组织
terry weave 机织毛圈组织

corduroy [ˈkɔːdərɔɪ]	灯芯绒
grey cloth	毛坯布
greige [greɪʒ] cloth	毛坯布
finished fabric	成品织物
visual weaving fault	外观织疵
tension	张力
machine setting	机器设置
malfunction [mælˈfʌŋkʃən]	故障
weft "cracking"	纬向条花
warp "streakiness"	经向条花
warp "smashing"	经纱崩纱
heald mark	综丝擦痕
reed mark	筘痕
production lot	生产批次
misdrawing	错穿
loom fly	织入飞花
oil stain	油污迹

Exercises

1. Which of the following description(s) is (are) correct? (　　)
   a) The more heald shafts a loom has, the more complex are the structures that the loom can produce.
   b) On a shuttle loom, it is not necessary for yarn packages to be transported though the shed when the weft is being inserted.
   c) Shuttleless looms usually use selvedge tuck-in devices or produce a fringed selvedge.
   d) The pick density within a fabric will be determined by tensions in the warp and the fabric, and the let-off and take-up rates, together with the linear density of the weft yarns.
2. The term for the action of pushing the weft yarn firmly against the fell of the

fabric by the reed is _______. ( )

a) pushing b) let-off c) picking d) beating up

3. The simplest woven structure is _______. ( )

a) twill b) sateen

c) plain weave d) leno weave

4. Which of the following woven structures is (are) derivative weave(s)? ( )

a) Sateen b) Oxford

c) Backed weave d) Herringbones

5. Grey cloth refers to _______. ( )

a) cloth in a grey colour

b) cloth being cut for clothing

c) coloured cloth

d) cloth just leaving the weaving machine

6. Weft "cracking" on the fabric may be caused by _______. ( )

a) insufficient tension in warp

b) malfunctioning of the weft insertion device

c) irregular weft linear density

d) burrs on the reed wires

7. Which of the following commercially used fabrics is a plain weave structure? ( )

a) Herringbone b) Khaki c) Habotai d) Serge

8. The structure shown in the diagram is _______. ( )

a) a complex structure

b) a twill derivative structure

c) a satin derivative structure

d) a plain derivative structure

9. The structure shown in the diagram is _______. ( )

a) a complex structure

b) a twill derivative structure

c) a satin derivative structure

d) a plain derivative structure

10. The structure shown in the diagram is ________. (   )

a) a twill derivative structure

b) a structure of sateen

c) a structure of gabardine

d) a structure of warp face satin

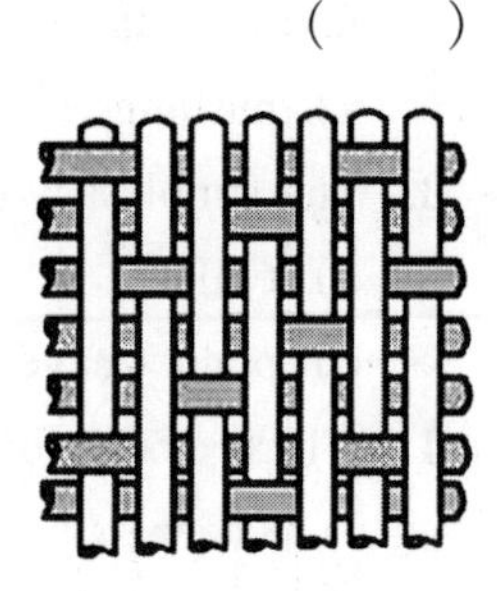

Reading Materials

Weaving Machine Developments

The star of the weaving room continues to advance, offering added value and higher quality to finished products.

TW Special Report

Even though, in most case, the weaving process is the same as it ever was — create a shed, insert the filling and beat up the pick — today's weaving machines are highly technical instruments featuring numerous innovations that have sped-up production, improved quality and saved energy.

Many of the latest developments in weaving machines have focused on automation, digital upgrades and innovative end products that add value including technical textiles, e-textiles and 3D structures.

Itema reports the Itematech A9500^2 air-jet weaving machine is especially welcomed in demanding weaving markets in the United States, Russia and India, among other countries. The machine combines two Itema technologies — for shed geometry and air-jet weft insertion — to guarantee high-quality fabrics even when weaving at high speeds.

— excerpted from *Textile World*, September/October 2021, page 18

【参考提示】

1. Itema，意大利著名的织机制造商，国内有译为"意达织造"的。Itematech 为该公司旗下专攻技术织物织造方面设备的公司。

2. A9500^2，Itematech 的喷气织机型号。

3. shed geometry，开口的几何形态。

Picanol Launches New Connect Generation Airjet and Rapier Weaving Machines

Picanol has introduced its latest generation of airjet and rapier weaving machines, which have been called the "Connect" generation. These new generation weaving machines focus on connectivity and an increased level of data availability. With this new generation, Picanol is launching several new functionalities such as a digitalized Gripper stroke setting, Gripper tape monitoring, Climate control, Shed angle measurement, and fully integrated Power monitoring.

"Following the successful launch of our digital platform PicConnect earlier this month, we now released a new generation of weaving machines, which are known as the Connect generation. Our Connect generation weaving machines can provide the correct data and are loaded with new and never-before-seen functionalities. This is clear proof that for Picanol 'Driven by Data' is not just a slogan but a commitment. Not only have these innovations allowed us to make big improvements when it comes to our four design principles — Smart Performance, Sustainability Inside, Intuitive Control, and, of course, Driven by Data — but when combined with our new digital platform they will allow you to amplify your own intuition." comments Johan Verstraete, Vice President Weaving Machines of Picanol.

From now on, the range of Picanol weaving machines will have the "Connect" suffix. For the airjet weaving machines, this means: OmniPlus-i Connect and TerryPlus-i Connect, while for the rapier weaving machines, this means: OptiMax-i Connect and TerryMax-i Connect. These new machines are built around Picanol's four design principles.

— excerpted from *International Textile Market*, November-December 2021, pages 27 & 28

【参考提示】

1. Picanol，总部在比利时的知名织机制造商，国内有译为"必佳乐"的。

2. gripper，本文中指"（剑杆的）剑头"。因此 gripper stroke setting 应指"剑头冲程设置"，而下文的 gripper tape，应该指"（挠性剑杆的）剑带"。

3. Climate control，指带温湿度传感器的"温湿度控制装置"。

Weaving: Auxiliary Equipment Matters

Weaving is a complex process requiring technology beyond the weaving machine for the highest-quality fabrics.

TW Special Report

Quality Control and Monitoring

The highest-quality yarns and equipment sometime fail. Possibly preventing the failure, or knowing when and where a failure occurred during weaving is necessary to fix problems to avoid downstream processing issues.

Switzerland-based Loepfe Brothers Ltd. offers weavers weft-control solutions for all weft insertion systems, as well as yarn break solutions for demanding weaving applications. The WeftMaster SFB weft thread brake controls the weft tension for all yarn types on rapier and projectile weaving machines. According to the company, an optimized, electronically controlled late braking start by the projectile sensor produces a uniform weft thread tension over the entire width of the fabric.

— excerpted from *Textile World*, September/October 2021, page 25

【参考提示】

1. Loepfe Brothers Ltd. ,瑞士知名的纺织电子仪器生产商,国内有译为“洛菲兄弟公司”的。
2. demanding... ,指“……(方面)日益增长的要求”。

CHAPTER 4

KNITTING AND KNITTED FABRICS

Knitted fabrics are classified as being either weft or warp knitted. Irrespective of this classification, all knitted fabrics are formed by interlacing loops of yarns. In the weft knitted fabrics each loop comprises two pillars, one needle loop and one sinker loop. In warp knitted fabrics the loop comprises two pillars, one needle loop and one underlap (see Figs. 4.1 and 4.2).

One widthwise row of loops is called a "course" and one lengthwise column of loops is called a "wale".

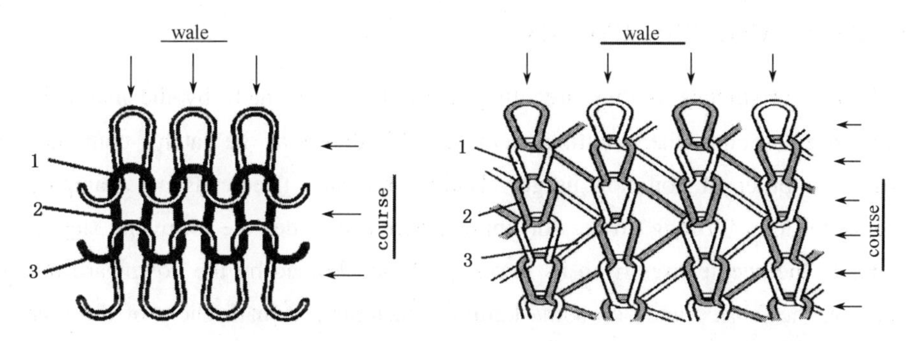

Fig. 4.1 Back side of plain knit (single weft knit)

1—needle loop 2—pillar 3—sinker loop

Fig. 4.2 Face side of atlas (single bar warp knit)

1—open loop 2—closed loop 3—underlap

In a single knit fabric (or single jersey), which is knitted on a single set of needles, the pillars and loops appear respectively in different sides of the fabric. The side showing pillars only is referred to as the technical face side, which is smoother because the alignment of the pillars allows a better reflection of light. The reverse side showing loops is called the technical back, which seems rougher due to the diffuse reflection of light on the loops. The term "technical" is used because in the final product produced from the fabric, the technical back may be used as the actual face side of the product.

If the fabric is knitted using two sets of needles, both the loops and pillars appear on one side of a knitted fabric, and such a knitted fabric is called as double knit (or a double jersey) fabric.

In weft knitting, each yarn is formed into a loop by needles in succession to produce a course. As these loops are intermeshed with successive loops knitted on the same needle, columns of loops (called wales) are produced.

In warp knitting, warps of yarn form loops simultaneously to produce a course of loops; therefore warping is necessary before knitting.

1 KNITTING PROCESS AND KNITTING ELEMENTS

An understanding of how the needles create the knitted loops is necessary in order to appreciate the different structures and properties of knitted fabrics.

1.1 BASIC KNITTING CYCLES

Loops, or stitches as they are often referred to, are made by the interaction of knitting needles and other knitting elements. Needles may be latch, compound or bearded, in order of common usage. Taking the latch needle as an example, the fundamental loop forming process or knitting action includes the following steps:

First, the needle rises from its lowest position. Forced by the downward take-up tension on the fabric, and, on some knitting machines, with the help of the holding-down function of the sinker throat, the old loop slides down from the needle hook, and opens the needle latch. As the needle continues to ascend, it passes over the latch spoon and onto the needle stem, to complete the so-called loop-clearing stage. The needle is now at its highest position (see Fig. 4. 3 from 1 to 5).

As the needle descends from its highest position, a new yarn is introduced into the needle hook. As the needle continues to descend, it passes through the loop on the needle stem and, as it does so, this "old" loop closes the latch and lands on top of it. Further downward movement causes the old loop to be knocked over the needle head, drawing the yarn in the hook of the needle into a loop shape, through the cast-off old loop to form a new loop. Under the take-up force, the newly formed loop is pulled to back side of the needle ready to become the old loop for the next loop (see Fig. 4. 3 from 5 to 11). As each needle knits in succession, a course of loops is formed.

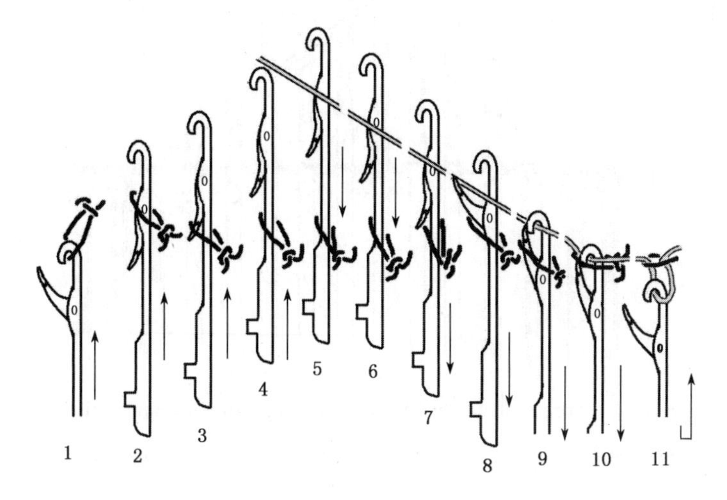

Fig. 4. 3 Illustration of loop-forming cycle in weft knitting with latch needles

On a circular weft knitting machine, the needles are inserted in vertical slots, called tricks, cut into a cylinder.

If one individual needle, holding an old loop, does not ascend from its lowest position to clear the old loop, the new yarn will not be laid in its hook. On that wale, the yarn will become a float in the course formed by that yarn and the old loop held by that needle will be extended (see Figs. 4. 4 and 4. 5. The former shows a float to be formed on needle "a"). This action is called as non-knitting, or missing and the stitch to be formed is called a miss stitch or float stitch.

If one individual needle only rises to a position where the old loop is still on the latch, the newly introduced yarn will not be drawn through the old loop but will be tucked in behind it. The loop formed is called a tuck loop or tuck stitch. Both the old loop and the tuck loop will be cast off the next time that the needle undergoes a full knitting cycle (see Fig. 4. 4 and 4. 6. Tucks will be formed on needles "b" and "c"). This action is called tucking.

It can be seen that one up-and-down movement of the needle fulfils one cycle of loop formation. In weft knitting, the up-and-down movement of the latch needle is caused by the relative movement between a butt which protrudes from the needle and a

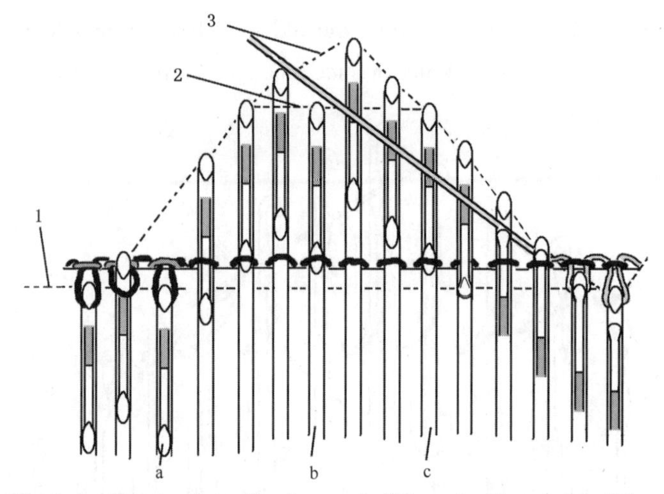

Fig. 4. 4 Traces of needles in non-knitting, tucking and knitting

1—trace of non-knitting needles 2—trace of tucking needles 3—trace of knitting needles

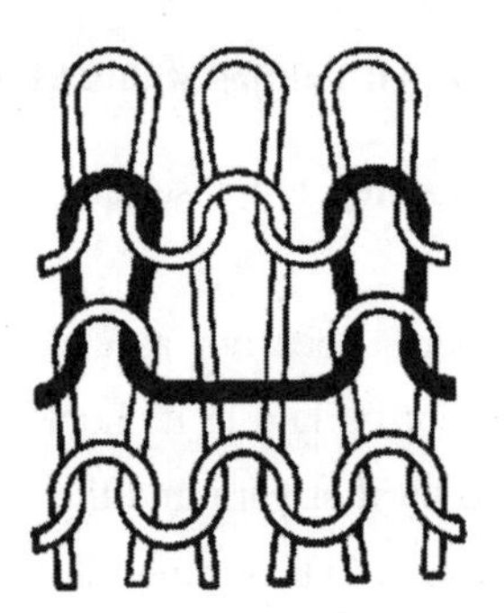

Fig. 4. 5 A single float stitch due to non-knitting

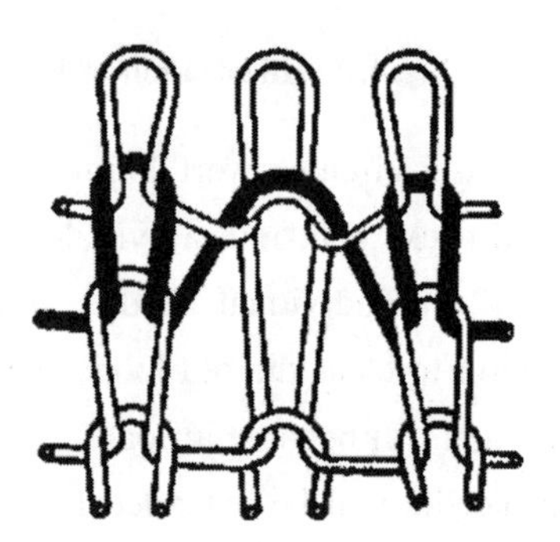

Fig. 4. 6 A single tuck stitch

cam track formed by a group of cams, such as a raising cam, clearing cam, knocking-over cam and stitching cam, etc. (See Fig. 4. 7).

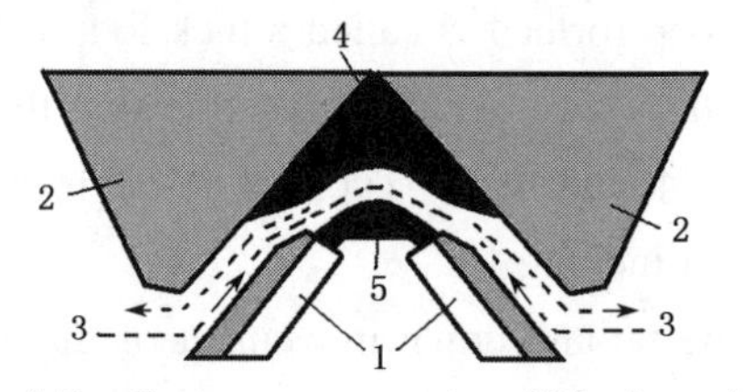

Fig. 4. 7 Cam arrangement on V-bed-machine

1—raising cams 2—stitching cam 3—cam track 4—depressing cam 5—clearing cam

On a circular weft knitting machine, a group of such cams with a yarn feeder or yarn carrier constitutes a feed or feeder, and the number of feeds on a machine determines the maximum number of the courses which could be knitted in one rotation of the needle cylinder. On most circular weft knitting machines, the cam boxes are stationary and the rotation of the needle cylinder forces the needles in the tricks on the cylinder through the cam track to move up-and-down (see Fig. 4.8).

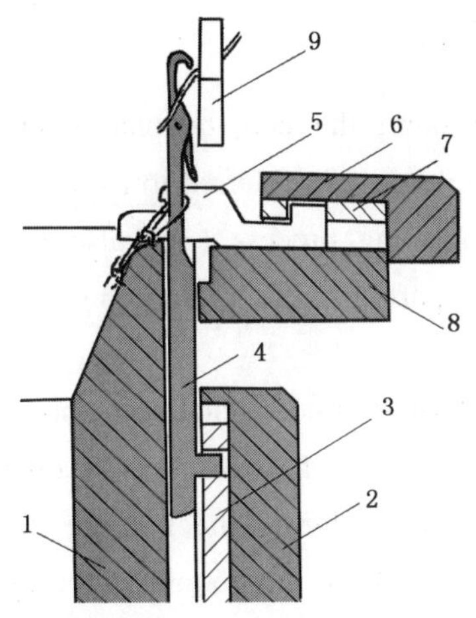

Fig. 4.8 Knitting elements on a circular single knitting machine

1—cylinder 2—cam box 3—cam 4—latch needle 5—sinker
6—sinker cam ring 7—sinker cam 8—sinker dial 9—yarn carrier (feeder)

On flat V-bed knitting machines (see Fig. 4.9), the needles are arranged in a straight line and inserted into tricks in a flat metal plate, and two sets of plates are aligned to form an inverted V shape. A cam box mounted on a carriage moves back and forth, causing the needles in the needle slots in both plates to move up and down (see also Fig. 4.7).

In warp knitting, all needles are fixed into a needle bar, and warp yarns are carried by guides mounted on guide bar(s). A guide bar can swing from the front of the machine to the back, and vice versa, and it can also

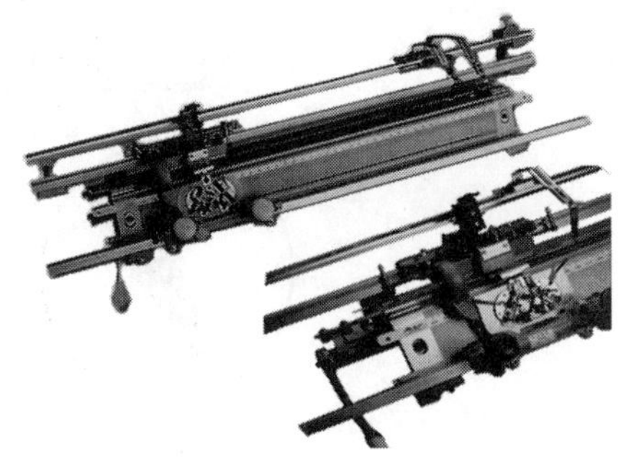

Fig. 4.9 Manually operated V-bed machines

move laterally, so that the guides pass between the needles to the needle front and then move laterally one needle space to wrap the yarns around them to make an overlap and, afterwards swing back between the needles and move laterally to create an underlap. Therefore one up-and-down movement of the needle bar and the lapping movements of the guides make one knitting cycle and forms one row (course) of loops simultaneously with hundreds of warp yarns.

1.2 KNITTING ELEMENTS

A fabric is knitted through the conjunctional interaction between a series of knitting elements, which are generally referred to as the machine parts in contact with the yarn during the loop forming process (see also Fig. 4. 8). A more detailed description of some of these parts is as follows.

1.2.1 Knitting Needles

Needles are one of the most fundamental elements on a knitting machine. The latch needle is a commonly used needle because it is opened or closed automatically under the action of the old loop during the knitting process. The latch needle is composed of the following parts from the bottom to the top (see Fig. 4. 10): shank, butt, stem, latch, hook and head. The latch mounted on a saw cut in the stem of the needle can pivot on either a rivet or localized indentation in the walls of the saw cut stem. The spoon of the latch can ensure the accurate closure of the needle hook and create a smoother profile for the old lop to pass over. The needles come in different sizes, defined by their gauge (or cut in the USA) which refers to the number of

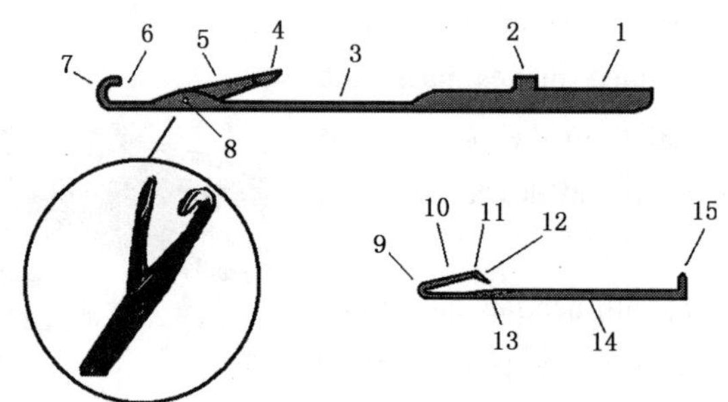

Fig. 4. 10 Latch needle and spring needle

1—shank 2—butt 3—stem 4—latch spoon 5—latch 6—hook 7—needle head
8—rivet 9—needle head 10—beard 11—bend 12—tip 13—eye 14—stem 15—butt

needles that may be accommodated per unit length (usually one inch) of the machine. The finer the gauge, the more needles per unit length, and the finer yarns that can be knitted, and hence, the lighter the weight of fabric that can be produced.

Faults in the performance of the latch would, almost inevitably, produce faults in the fabric. The needle butt, actuated upon by the cams, moves along the cam track. Some circular knitting machines have more than one cam track at each feed and each track corresponds to and aligns with a different level of needle butt. Thus, more than one type of knitting action can be performed on a yarn by successive needles. By setting the cam tracks to create either knit, tuck or miss stitches (by raising the needles to different heights or not at all, in the case of the miss stitch), and arranging the needles with different butt levels into a sequence, patterned fabric with a combinations of knit, tuck and miss stitches could be produced.

For double jersey circular weft knitting, a machine with two sets of needles are required, one in a cylinder and the other in a dial. If the cylinder needles align with the dial needle walls and vice versa, the needle set-out is called as "rib gating". This is used to produce rib gated structures.

Another gating arrangement is "interlock gating". Interlock gated machines use two sets of needles in both the dial and cylinder. Each set of needles has different levels of needle butt, either high-butt or low-butt level, and the cam boxes for both the cylinder and dial have two tracks, one for high-butt needles and the other for low-butt needles. The cylinder high-butt needles align with the dial low-butt needles and the cylinder low-butt needles align with the dial high-butt needles.

The spring or bearded needle is another traditional needle, which used to be used to make a very fine fabric since the bearded needle could be made finer than other types of needles. The bearded needle has a head, beard, bend, tip, eye, stem and butt (see Fig. 4. 10). The butt of the bearded needle is used to fix each needle into a needle bar. Alternatively, for convenience sake, several needles can be cast into a needle lead, which is then fixed into the needle bar. In order to close the beard of the needle to make the old loop land on the outside of it during the knitting cycle, a presser has to be used to press on the needle bend, forcing the needle tip into the needle eye and to close completely the beard. A cut presser may also be used for certain courses, in which case, selected needles are not closed and the old loop will pass into their hooks to form tuck loops.

As the machine speeds increased, the friction between needles and the presser caused larger heat generation and noise, which precluded the machine speed from further increase. Consequently, most machines that were equipped with bearded needles have been superseded by those using compound needles.

Nowadays the compound needle is widely used, especially on modern warp knitting machines. A compound needle is composed of two parts, a hook and a sliding or closing element. The relative movements of the two parts accomplish the knitting actions, opening and closing of the hook, and in this way the vertical movement or stroke of the needle can be reduced. This allows faster knitting speeds at the expense of more complicated drive mechanisms to control the movement of the two separate knitting elements (see Fig. 4.11).

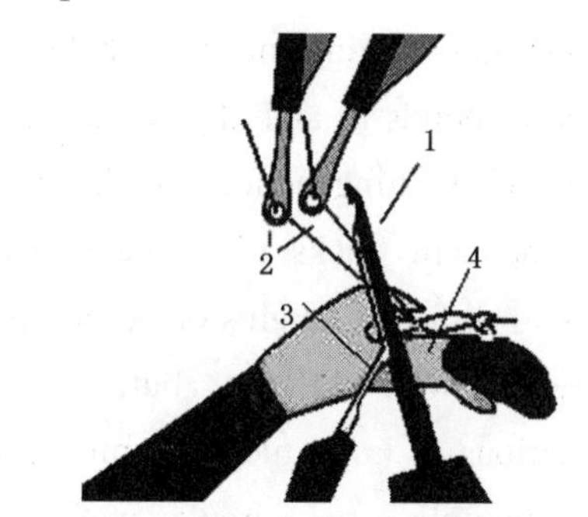

Fig. 4.11 Knitting elements with compound needle (warp knitting)

1—needle 2—guides 3—slider 4—sinker

1.2.2 Sinkers

On a single jersey machine sinkers are used (see Fig. 4.8). The sinker is designed to have a sinker throat that can hold the old loop during clearing to restrain the old loop as the needle rises. The sinker also has a belly which can support the sinker loop, and help the old loop to be cast-off and the new loop to be formed when the needle descends towards its lowest position.

A specially designed sinker nose in circular weft terry knitting machines enables a longer sinker loop to be formed to give a terry effect.

On most latch needle warp knitting machines, the shape of the sinker is much simpler and no throat or no nose is needed, since only the bottom edge of the sinker will be used to hold the fabric as the needle ascends; therefore such sinker is called a holding down sinker.

1.2.3 Yarn Feeders

In weft knitting, a yarn feeder (see Fig. 4.8) presents the yarn in such a position that it may pass into the hook as the needle descends. The shape of the yarn feeders may differ from one machine type to another due to the functions they perform. On basic circular latch needle machines, the yarn feeder is in a "boot" shape, the bottom of which acts as a latch guard to prevent the needle latch from swinging back to close the hook (and prevent the insertion of the new yarn into the hook) as the old loop clears off the latch.

If two yarns are knitted together at the same feeder to produce a plating effect, there would be two yarn-guiding holes on the feeder, one for the face yarn and the other for the backing yarn, and the position of these holes should present the yarns at different laid-in angles to the needles so that the face yarn will always appear on the face side of the fabric.

In warp knitting, guides are used, and each warp yarn should be threaded through the eye in a separate guide. Several guides are cast into a lead (see Fig. 4.12. Knitting elements on warp knitting machines are usually cast into lead since the same knitting elements on the machine always work simultaneously), which is assembled onto the guide bar. When the guide bar swings from the back of the needles to the front, or vice versa, and then moves laterally, the guides swing between the needle spaces and to make an overlap (to lay the yarns into the needle hooks) or make an underlap respectively. The lateral movement of the guide bars is controlled by cams, pattern chains with replaceable links of different height, or electro-mechanical actuating motors. Each guide bar controls a set of warp yarns, and its lapping movement can be controlled separately. The more guide bars a warp knitting machine has, the more lapping movements it could have for each course (see Figs. 4.11 and 4.13).

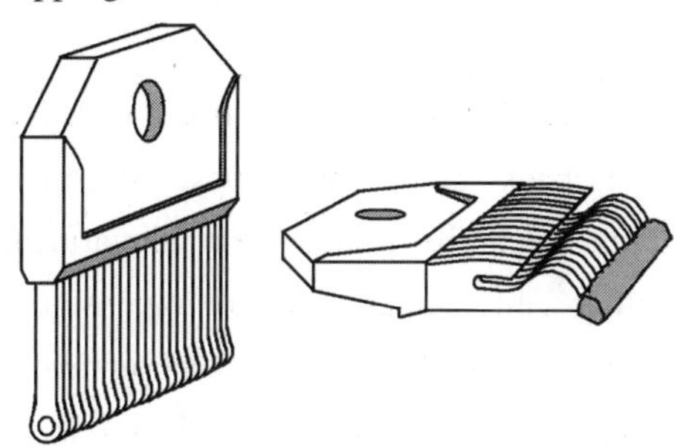

Fig. 4.12 Guides and sinkers set in lead

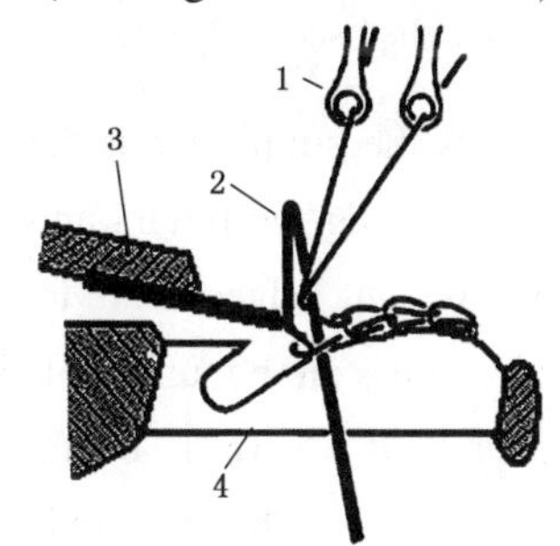

Fig. 4.13 Knitting elements with spring or bearded needle (warp knitting)

1—guides 2—spring needle

3—presser bar 4—sinker

2 KNITTED FABRICS

Owing to the different ways in which fabrics are formed, the characteristics of weft knits and warp knits are quite different in several respects.

2.1 WEFT KNITTED FABRICS

The primary structures for weft knitted fabrics are plain knit (single jersey, see Fig. 4.1), rib (Fig. 4.14), purl (links and links stitches, see Fig. 4.15) and interlock (Fig. 4.16).

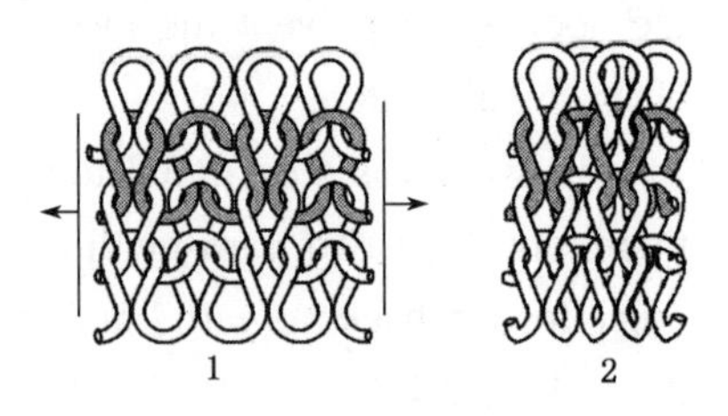

Fig. 4.14 1 × 1 rib

1—stretched state 2—relaxed state

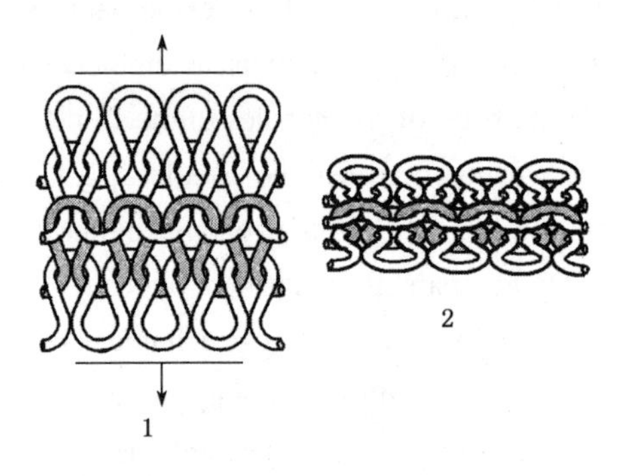

Fig. 4.15 Purl stitch

1—stretched state 2—relaxed state

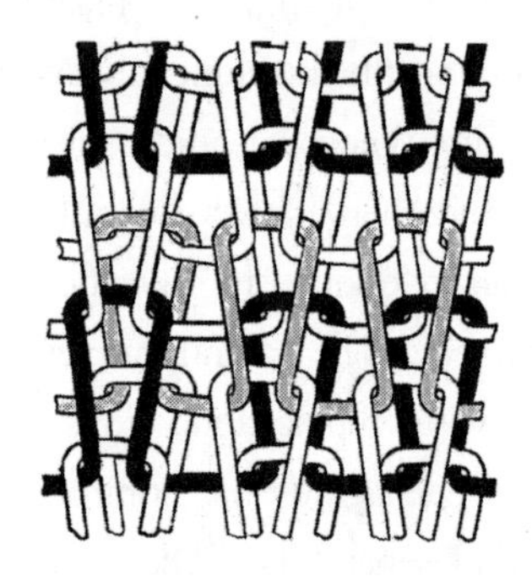

Fig. 4.16 Interlock

Plain knit is the simplest but the most widely used single jersey fabric, with the technical face showing the pillars and the technical back showing the loops. When cut into a piece, the cut edges tend to curl, which might be troublesome in garment making. Plain knit can be unravelled from either end, either along or against the knitting direction. If a yarn in a certain loop is broken, the loops in the wale unravel, and laddering occurs. The widthwise or coursewise extensibility of a plain knit is greater than that in the lengthwise or wale direction.

Rib is widely used in cuffs and necks for knitwear because it has a good elasticity and recovery in the course direction after the stretching force is released. Balanced rib structures are double jersey structures that are characterized by showing the same appearance on both sides. Commonly used rib structures are 1 × 1 rib (which means that the fabric has one wale of face loops and one wale of back loops alternatively) and 2 × 2 rib. A rib structure can only be unravelled from the end knitted last (against the knitting direction) and only the cut edges along the wale direction tend to curl.

Purl stitches can only be knitted on a special machine with double headed needles, and, because the loops are formed in successive courses in alternate needle heads, the face loop course and the back loop course will alternatively appear on either side of the fabric. The purl structure has a good walewise recovery after extension. Circular purl machines have two cylinders superimposed one on top of the other. They can also produce double jersey structures and are commonly used in double cylinder hosiery knitting.

Interlock is actually a combination of two 1 × 1 rib structures, one formed by the high-butt needles and the other formed by the low-butt needles. Interlock is suitable for winter underwear since it has a good heat retention property. Like the rib stitches, interlock can only be unravelled against the knitting direction but has less extensibility than 1 × 1 rib in both directions.

There are many fancy structures based on the above primary structures in weft knitting. The jacquard structure is one of the common fancy structures. A single jersey jacquard structure is generally achieved by the combination of knit, miss-knit and tuck stitches using mechanical or electro-mechanical needle selection mechanisms, and a double jersey jacquard is usually achieved on the cylinder with the help of various needle selection mechanisms, which have traditionally used mechanical pattern drums, intermediate jacks and selection jacks.

Nowadays electronic selection devices are widely used. A needle selection system based on piezoelectric effects is one example. In the piezo system a needle selector is composed of two ceramic plates stuck to each other. The principle of the piezo (piezoelectric) needle selection is that a ceramic plate bends when an electric current is passing through it. The current may be applied in pulses according to signals corresponding to the desired pattern of stitches. The slight movement of the ceramic plates is amplified through levers in the selector. The movement of the levers deflect other knitting elements when they contact them and thereby execute needle selection. Electronic needle selection can produce very large pattern repeat areas.

Single jersey laid-in stitches and terry stitches are also common fancy structures. In the former, float yarns are held by loops formed by a face yarn and a ground yarn knitted using a plating technique, and in the latter, the terry loops are usually formed with the help of a special sinker nose.

Rib and interlock derivative structures may be produced using combinations of courses of knit, non-knit and tuck stitches on a rib or interlock base. These include Ponte de Roma, full cardigan, half cardigan, French double pique and Swiss double pique.

If special techniques and devices are used, more fancy structures could be achieved. For example, by racking the needle beds in a V-bed knitting machine or transferring loops, ripple or cable stitches can be made. By using a striper mechanism, a wide variety of coloured stripe patterns can be created, and by transferring needle loops from some needles to their adjacent needles, eyelet stitches can be knitted. Through stitch-transfer techniques, widening or narrowing can be achieved, and thus, to produce seamless garments directly from a knitting machine becomes possible.

2.2 WARP KNITTED FABRICS

Warp knitting machines can be classified as either Tricot or Raschel. The Tricot machine is usually a fine gauge machine with 2 to 4 guide bars that runs at a much higher speed and, and the fabric take-up direction on such a machine is nearly perpendicular to the needle plane.

The Raschel machine usually has many more guide bars, which allow more complicated structures to be produced because the lateral movement of each guide bar is controlled by an individual set of chain links on a pattern wheel or by an individual electronic actuator, and the fabric take-up direction on a Raschel machine is downwards at an angle of about 140 degrees to the needle plane.

Most warp knitting machines have a single needle bar; therefore only single knit fabrics can be produced, where loop pillars appear on one side (the technical face side) and the underlaps always appear on the other side (the technical back side). There are some warp knitting machines with double needle bars and they are usually used to knit pile fabrics used for automobile upholstery, spacer fabrics or tubular products such as net sacks.

The primary warp knitted structures are single bar chain stitches (Fig. 4.17), tricot (Fig. 4. 18) and atlas (see Fig. 4. 2). Chain stitch structures are characterized by one warp always forming loops on the same needle and yarns inserted by the other guide bars join these chains of loops together across the width of the fabric to make it coherent. Chain stitches have lower lengthwise extensibility and they are often used with other structures to present stability in the wale direction.

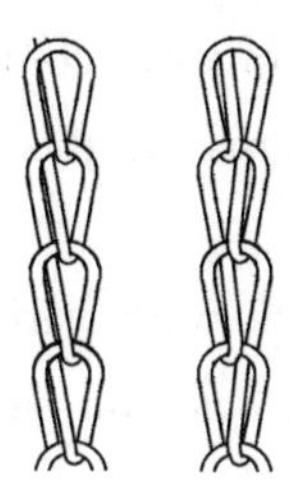

Fig. 4. 17 Chain stitch

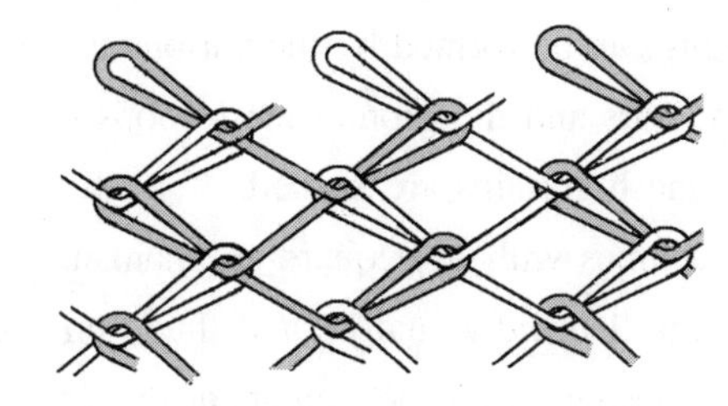

Fig. 4. 18 Tricot

For tricot stitches, each warp will form loops alternately on two neighbouring needles. The warp makes at least one needle space underlap after a loop is formed on one needle, and thus the next loop can be formed on an adjacent needle, and all loops can be linked together to form a fabric.

For atlas structures, the guides carrying warps make underlaps over one needle space in the same direction for a number of courses in succession, and then make the same sequence of movements in the reverse direction.

If the underlap is made over more than one needle space each time for the tricot or atlas, the extensibility or rigidity of the fabric can be varied as well as the appearance of the fabric. Generally speaking, the longer the underlap, the greater the fabric walewise extensibility will be.

In practice, single guide bar warp knitted fabrics are rarely used, and instead, two or more guide bars are used to make more stable fabrics, such as locknit (see Fig. 4. 19), reverse locknit, sharkskin, queenscord, etc. For structures with longer underlaps floating on the back, the back side underlaps can be raised or brushed to give a velvet effect. Since the underlap formed by the front guide bar is always superimposed on the underlaps formed by other guide bars, the underlap on the front guide bar determines the elasticity

Fig. 4. 19 Double guide bar fabric (locknit)

of the fabric. The shorter it is, the more stable the fabric structure is.

With partial or part set threading, various double guide bar net structures (open knits) can be formed, which is one of the great advantages of warp knitting compared with other fabric forming systems.

Various laid-in designs can also be formed on the base of the mesh structures. The designs can be formed by the various traces of the laid-in yarns controlled by their own guide bars and the ground mesh loops to hold those yarns. In this way, laces or patterned mesh curtains are knitted.

If machines with a jacquard mechanism are used, each warp yarn to be laid-in can be controlled individually, and thus, on a chain stitch base the pattern repeat of the laid-in designs could be tremendously large.

With addition of a yarn insertion mechanism, yarns can be inserted horizontally, vertically, or even in multi-axial directions. Because the yarns are inserted without being formed into loops, thick and stiff yarns can be used, which gives the fabric very high tensile strength in the inserting direction.

3 QUALITY ISSUES IN KNITTED FABRICS

The loop length is one of the most important factors affecting the fabric quality, and control of the loop length is necessary to ensure a good quality of the fabrics. On many weft knitting machines positive yarn feeding devices are used to ensure uniform loop length, and on most tricot warp knitting machines, a positive let-off mechanism is also used for the same purpose. However, for Raschel machines to knit large patterns, negative let-off mechanisms have to be used since the warp run-in for each guide bars is quite different from course to course.

It can be seen from the above that all loops in a wale are formed by the same needle, and if some needles do not function correctly, faults would appear on the corresponding wales. Faults that appear in vertical lines in a fabric are usually due to the needles. In warp knitting, if one sectional beam is not set correctly, the run-in tension of the warps from that beam would be different from that of other warps, which may cause streakiness.

In weft knitting, if the setting of the stitch cams at some feeds differs from those at other feeds, or if the yarn tension is set incorrectly, or the apparent quality of the yarn being fed by any feeder is outside the tolerances set, horizontal lines called barre may appear in the fabric.

In warp knitting, horizontal lines are usually caused by uneven let-off tension, or uneven take-up tension, or even by any stop-start action on a high speed warp knitting machine.

In addition to incorrect machine settings, defective knitting elements would cause holes in fabrics, or cause unwanted tuck or dropped stitches. The temperature and humidity in the workshop could also cause knitting problems especially when synthetic yarns are being knitted on a fine gauge machines.

Words and Phrases

knitting and knitted fabrics	针织和针织物
loop	线圈
pillar	圈柱
needle loop	针编弧
sinker loop	沉降弧
underlap	延展线
course	横列
wale	纵行
plain knit	平针
atlas [ˈætləs]	经缎
single weft knit	单面纬编
single bar warp knit	单梳经编
single knit	单面针织
single jersey [ˈdʒɜːzɪ]	单面针织物
needle	织针
technical face	工艺正面
technical back	工艺反面
double knit	双面针织
double jersey	双面针织物

intermesh [ˌɪntəˈmeʃ]	串套
stitch	(针织)组织,线圈
knitting element	成圈机件
latch needle	舌针
loop forming process	成圈过程
knitting action	编织动作
sinker throat	沉降片喉
old loop	旧线圈
clearing	退圈
land on	套圈
landing	套圈
knock over	脱圈
needle head	针头
cast-off	脱圈
circular weft knitting machine	圆纬机
trick	针槽
cylinder	针筒
float	浮线
non-knitting	不编织
missing	不编织
miss stitch	浮线组织
float stitch	浮线组织
tuck loop	集圈
tuck stitch	集圈组织
tucking	集圈
cam track	三角跑道
raising cam	起针三角
clearing cam	退圈三角
knocking-over cam	脱圈三角
stitching cam	弯纱三角
yarn feeder	导纱器
yarn carrier	导纱器

feed	成圈系统(成圈路数)
flat V-bed knitting machine	横机
cam box	三角座
carriage	(横机)机头/游架
depressing cam	压针三角
cam arrangement	三角配置
needle bar	针床
guide	导纱针
guide bar	梳栉
overlap	针前垫纱
underlap	针背垫纱
sinker	沉降片
sinker cam ring	沉降片三角环
sinker cam	沉降片三角
sinker dial	沉降片盘
shank	(针)尾,(针)杆
butt	(针)踵
stem	(针)杆
latch	(针)舌
hook	(针)钩
saw cut	针舌槽
rivet [ˈrɪvɪt]	针舌销
localized indentation	定位凹点
latch spoon	针舌勺
gauge [geɪdʒ]	机号
cut (*Ame.*)	机号
patterned fabric	带花纹的织物
beard [bɪəd]	(钩针)针钩
bend	(钩针)针鼻
tip	(钩针)针尖
dial	针盘
dial needle	针盘针

rib gating	罗纹对针
rib gated structure	罗纹对针的结构
interlock [ˌɪntəˈlɒk] gating	双罗纹对针
high-butt needle	高踵针
low-butt needle	低踵针
spring or bearded needle	弹簧针/钩针
presser	压板
cut presser	花压板
compound needle	复合针
hook	针钩
sliding or closing element	针芯
sinker belly	沉降片片腹
sinker nose	沉降片片鼻
circular weft terry knitting machine	圆型纬编毛圈机
latch needle warp knitting machine	舌针经编机
holding down sinker	握持沉降片
latch guard	护针舌器
plating effect	添纱效应
yarn-guiding hole	导纱孔
laid-in angle	垫纱角
face yarn	面纱
lateral [ˈlætərəl] movement	横移
cam	凸轮,三角
pattern chain	花板链
pattern link	花板
guide needle lead [led]	导纱针蜡
sinker lead	沉降片蜡
slider	(复合针)针芯
presser bar	压板床
rib	罗纹
purl [pɜːl] stitch	双反面组织
links and links stitch	双反面组织

interlock	双罗纹
1 ×1 rib	1 +1 罗纹
curl	卷边
unravel [ʌnˈrævəl]	脱散
along the knitting direction	沿编织方向
against the knitting direction	逆编织方向
laddering	梯脱
cuff	袖口
neck	领口
knitwear [ˈnɪtweə]	针织服装
balanced rib structure	平衡的罗纹结构(相同数量的正、反面线圈纵行交替配置的罗纹结构)
2 ×2 rib	2 +2 罗纹
double headed needle	双头织针
double cylinder hosiery knitting	双针筒袜类编织
heat retention [rɪˈtenʃən]	保暖
fancy structure	花式组织
needle selection mechanism	选针机构
pattern drum	提花滚筒
intermediate [ˌɪntəˈmiːdjət] jack	中间推片
selection jack	选针片
piezo [paɪˈiːzəʊ] system	压电系统
piezoelectric [paɪˌiːzəʊɪˈlektrɪk]	压电的
needle selector	选针器
ceramic plate	陶瓷盘
electric current	电流
electric voltage [ˈvəʊltɪdʒ]	电压
single jersey laid-in stitch	单面衬垫组织
terry stitch	毛圈组织
ground yarn	地纱
plating techniques	添纱技术
rib derivative structure	罗纹变化组织

interlock derivative structure	双罗纹变化组织
Ponte de Roma	蓬托地罗马双罗纹空气层组织
full cardigan [ˈkɑːdɪgən]	全畦编
half cardigan	半畦编
French double pique [ˈpiːkeɪ]	法式点纹
Swiss double pique	瑞士式点纹
racking	针床横移
cable stitch	绞花组织
ripple stitch	波纹组织
eyelet [ˈaɪlɪt] stitch	纱罗组织
striper mechanism	调线机构
widening	放针
narrowing	收针
seamless garments	无缝成型服装
warp knitting machine	经编机
Tricot [ˈtrɪkəʊ] machine	特利考经编机
Raschel [rɑːˈʃel, rə-] machine	拉歇尔经编机
pattern wheel	花板轮
single needle bar	单针床
double needle-bar	双针床
pile fabric	割绒织物
automobile [ˌɔːtəməˈbiːl] upholstery	汽车内装潢
spacer [ˈspeɪsə] fabric	间隔织物
tubular [ˈtjuːbjʊlə] product	圆筒产品
net sack	网袋
chain stitch	编链
tricot	经平
needle space	针距
double guide bar fabric	双梳织物
locknit [ˈlɒknɪt]	经平绒
reverse locknit	经绒平
sharkskin [ˈʃɑːkskɪn]	经斜平

queenscord	经斜编链
raised	经起绒的
brushed	经刷毛的
velvet [ˈvelvɪt]	天鹅绒
front guide bar	前梳栉
partial threading	部分穿经
part set threading	部分穿经
net structures	网眼结构
open knits	网眼针织物
laid-in design	衬垫花型
mesh structure	网眼结构
laid-in yarn	衬垫纱
lace	花边
patterned mesh curtain	带花纹的网眼结构窗帘
yarn insertion mechanism	衬纱机构
multi-axial directions	多轴向
loop length	线圈长度
positive yarn feeding device	积极喂纱装置
positive let-off mechanism	积极送经装置
negative let-off mechanism	消极送经装置
warp run-in	经纱送经量
vertical line	纵向条花
sectional beam	分段经轴
streakiness	条花
barre [ˈbɑːreɪ]	纬向条花
horizontal line	水平条花
tuck	花针
dropped stitch	漏针

Exercises

1. Which of the structure(s) shown below can not be knitted on a V-bed weft

knitting machine? ( )

a) 1 ×1 rib b) Purl structure

c) Single jersey d) Tricot

2. Which word is most related to the part of the knitting action shown in the diagram? ( )

a) Clearing

b) Landing

c) Raising

d) Knocking-over

3. What type of stitch will be formed on the needle shown in the diagram? ( )

a) A normal knitted stitch

b) A float stitch

c) A tuck stitch

d) A miss stitch

4. Which of the following structures fall (s) into the category of double knit structures? ( )

a) Plain knit b) Interlock

c) Locknit d) 2 ×2 rib

5. Faults that appear as vertical lines in a weft knitted fabric are usually due to ________. ( )

a) abnormal yarn tension

b) malfunction of the corresponding knitting needle(s)

c) improper setting of the stitch cam

d) faults in the weft yarn

6. Which type of knitting needle given below can open or close its hook automatically during the knitting action? ( )

a) A compound needle b) A spring needle

c) A latch needle d) A bearded needle

7. Which of the following elements is needed in the formation of weft knitted Terry fabric? ( )

a) A striper mechanism b) A sinker with a special sinker nose

c) A double headed needle d) Guide bars

8. The setting of which of the following cams will affect the loop length? ( )
   a) Raising cam b) Clearing cam
   c) Tucking cam d) Stitching cam
9. The structure shown in the diagram is ________. ( )
   a) a single jersey weft knitted structure
   b) a double jersey weft knitted structure
   c) a single guide bar warp knitted structure
   d) a double guide bar warp knitted structure
10. For 1 × 1 rib structure, which of the following statements are false? ( )
   a) 1 × 1 rib structures can be produced on a single jersey machine.
   b) When cut into a piece, the cut edges of 1 × 1 rib fabrics along the course and wale directions tend to curl.
   c) 1 × 1 rib structures can be unravelled along the knitting direction.
   d) 1 × 1 rib structures often used in cuffs for knitwear due to its good coursewise elasticity.

Reading Materials

Karl Mayer Adds TM 4-S to Four-bar Tricot Machine Range

Karl Mayer expanded its four-bar tricot machine range with the TM 4-S. The TM 4-S model features a maximum speed of 1,800 rpm and a working width of 280″ that can be further extended by 10″. The needle and guide bars were manufactured with carbon fiber technology for high temperature stability and low weight. The machine is suitable for a variety of materials, such as silk net, raised velour fabrics, shoe fabrics, home textiles or upholstery for the automotive sector.

The TM 4-S is available in gauges E 28 and E 32 and standard equipment includes a 4 × 32″ beam frame.

— excerpted from *International Fiber Journal*, Issue 5 2021, page 11

【参考提示】

1. rpm, 即 revolutions per minute,每分钟转速。
2. E 28 和 E 32 为经编机机号,表示针床上"每英寸 28 针"和"每英寸 32 针"。
3. 4 × 32″ beam frame,能放置 4 个 32 英寸分段经轴的经轴架。

Karl Mayer Introduces Digital Solutions with Its HKS 3-M ON

Karl Mayer launched its HKS 3-M ON, featuring the use of digitally generated tricot machine which works with electronic pattern data obtained from the KM. ON cloud. This allows it to offer short and flexible design changes, minimal downtime and operation without the expense of ordering pattern discs. This also reduces the risk of operating errors.

The HKS 3-M ON also features the Spring Motion Assistant, an automatic return device that simplifies the changing of guide bars. There is a function on the touchscreen that guides the user through the partially automated process step by step.

— excerpted from *International Fiber Journal*, Issue 1 2022, page 12

【参考提示】

1. KM. ON cloud, 卡尔迈耶提供的安全云端服务。
2. design changes, 花型变化。
3. downtime, 停车时间;pattern discs,花盘。
4. Spring Motion Assistant,弹簧辅助装置;return device,复位装置。

Knitting Technology Developments

Knitting innovations continue to add value for textile manufacturers during a difficult time.

TW Special Report

Courses, wales, loops, gauge, warp, weft, raschel, jersey, interlock ... just some of the terms familiar to people working in the knitting industry. Knitting technology comes in a variety of types and sizes and may be used to make all sorts of products from small vascular heart grafts to shoe uppers and apparel fabrics all the way to large-scale bedding components and anything in between that requires comfort, stretch and seamless shaping, among other attributes.

Mayer & Cie

After completing field tests, Germany-based Mayer & Cie reports its OVJA 2.4 EM circular knitting machine now is ready for the market. The fully electronic model is designed for mattress cover fabrics with high output and a wide pattern variety. This ready-for-market version builds on the model introduced during ITMA 2019 with the addition of a new thread fluctuation control system positioned on every second feeder. A controlled air stream maintains constant yarn tension especially at high revolutions per minute, which helps avoid thread loops and dropped stitches.

Mayer & Cie also offers the OVJA 1.6 EE/2 WT double jersey jacquard machine for multi-colored designs and microstructure elements including tuck structures, spacer fabrics and double-knit fabric with a lay-in thread. The machine features three-way electronic needle selection in the cylinder and two-way technology in the rib dial. When special needles are employed, the machine can knit yarns up to 1,200 denier in gauges as coarse as E16. Applications include mattress covers, upholstery, transportation seat covers, shoe uppers and outerwear.

— excerpted from *Textile World*, January/February 2021, page 24

【参考提示】

1. ... just some of the terms familiar to people,严格说,just 前应该加 are。

2. vascular heart grafts,移植用心血管(人造心血管)。

3. shoe upper,鞋帮。

4. Mayer & Cie,德国知名纬编针织圆机制造商,国内有译为“迈耶·西”的。

5. ... thread loops and dropped stitches,这里的 thread 不会是工艺意义上的两纱加捻形成的“线”,纬编大圆机一般使用纱或长丝。根据上文意思,thread loops 应指垫纱张力波动引起的“纱(异常)起圈”。

6. two-way,three-way,即所谓的“两功位,三功位”,指通过选针装置、三角配置或高低踵织针配置来形成“编织成圈、浮线和集圈”中的两种或三种织物结构。

CHAPTER 5

NON-WOVEN FABRICS

1 TYPES OF NON-WOVEN FABRICS

Non-woven fabrics are by no means new types of fabrics. The concept of using heat, moisture and pressure to entangle wool or animal hair together to produce a piece of felt has been known for centuries. Nowadays modern technology provides many approaches to produce non-woven fabrics for different usages.

According to their structures, the non-woven fabrics can be classified as follows.

1.1 WEB-BASED NON-WOVEN FABRICS

For the web-based non-woven fabrics, fibre webs or batts should first be prepared through a technology of carding, wetlaying or airlaying, etc., and then by mechanical, chemical, thermal or solvent means, the fibres in the web are held together to form a fabric. When preparing the webs, fibres could be arranged multi-directionally or randomly to give all round stability and strength, or they can be arranged to be more aligned, usually parallel to the machine direction, to ensure the lengthwise stability and strength, or they can also be laid diagonally in a cross-machine direction. Composite webs can also be engineered whereby two or three webs, produced with different fibres, or in different orientations, or through different web-forming systems etc., are laid on top of each other and then bonded to produce a structure with the desired properties.

1.1.1 Fibre Bonding

In bonded non-woven fabrics, after web formation using either continuous filaments or staple fibres, chemical agents or heat is applied to bond the fibres together.

Acrylic or acrylate bonding-agent is one of the commonly used chemical agents. This may be applied as a powder, foam or by immersing the web in a liquid to saturation to bond it (see Fig. 5.1). The web is then passed over a heat cylinder or through a hot airflow oven to cure the chemical and, if necessary, dry the web. Non-woven fabrics formed in this way have good porosity but are somewhat stiff in handle. Typical applications are wall coverings and disposable products.

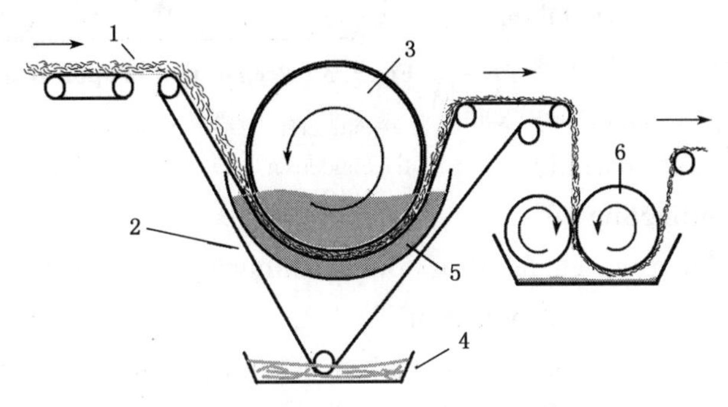

Fig. 5.1 An example of applying bonding agent

1—web 2—endless delivery apron 3—rotary screen drum
4—cleaning trough for delivery apron 5—immersion trough 6—squeezing roller

Heat bonding is the other method of bonding the web of fibres together. Thermo-fusible fibres such as polyamide should be used. The application of heat is usually done by passing the web between or around heated rollers (see Fig. 5.2), some of which may be engraved to emboss a pattern into the surface of the bonded web. This causes the fibres in the webs to soften and stick together, and, on cooling, bonds are formed within the fibre structure. Non-woven fabrics formed in this way are generally bulky with better filtration, porosity and resilience properties. They can be used as wadding in a quilt or winter clothing, filter cloths, or the base cloth for a tufted rug or carpet.

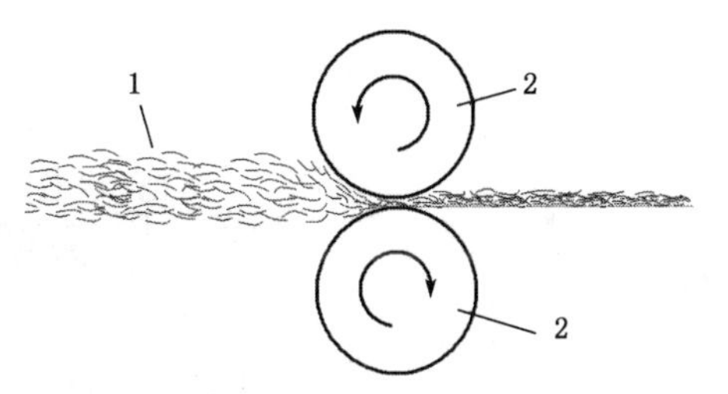

Fig. 5.2 Illustration of heat bonding

1—web 2—heated rollers

Spun-bonded fabrics also fall into this type. When the spinning solution is extruded from the spinneret, static electricity and high-pressure airflow are applied, and the fibres are laid randomly to form a laminate (see Fig. 5. 3). After being heat set by passing the web over a hot cylinder or being reinforced by needle punching, the non-woven fabric is made. Spun-bonded fabrics are used in agriculture and they can also be used as carpet backing.

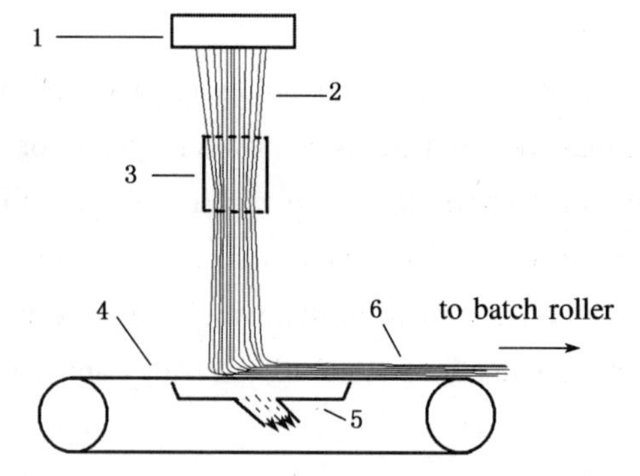

Fig. 5. 3 Preparation of spun-bonding laminate

1—spinneret 2—cooling area 3—drafting device 4—condenser screen 5—suction device 6—web

1. 1. 2 Fibre Entangling

This type of non-woven fabric is formed through a mechanical means, such as needle punching and hydroentanglement.

In needle bonding, thousands of barbed needles punch through the fibre webs, forcing fibres from one side through the web to entangle and lock them together to form the non-woven fabric (see Fig. 5. 4).

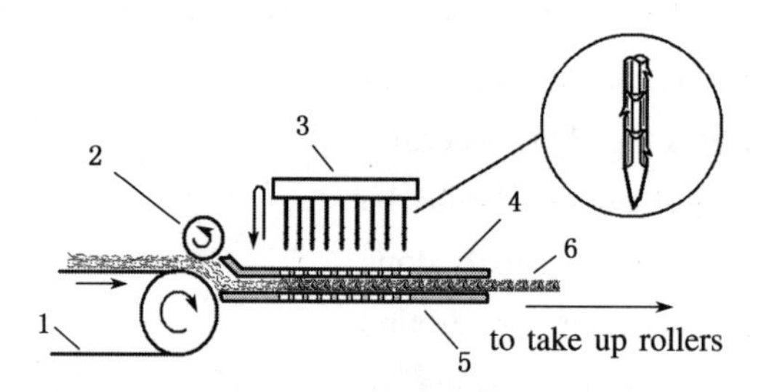

Fig. 5. 4 Illustration of needle punching

1—delivery apron 2—compression roller 3—barbed needles 4—compression plate 5—web plate 6—needled fabric

Needle punching, or needle felting as it is called, may be applied to many fibre types particularly wool. Needled fabrics are widely used as geotextiles, filtering materials, base cloths for artificial leathers and carpet tiles.

Hydroentanglement works on a similar principle but uses fine jets of water under high pressure instead of barbed needles (see Fig. 5. 5). The water jets penetrate the fibre web, prepared as either dry-or wet-laid webs, and cause the fibres to entangle.

When the water jets encounter the delivery apron or cylinder supporting the web, they rebound. This causes further fibre entanglement in the web. Fabrics produced using this method are often referred to as spun-lace fabrics. The characteristic properties of these fabrics are softness and drapeability.

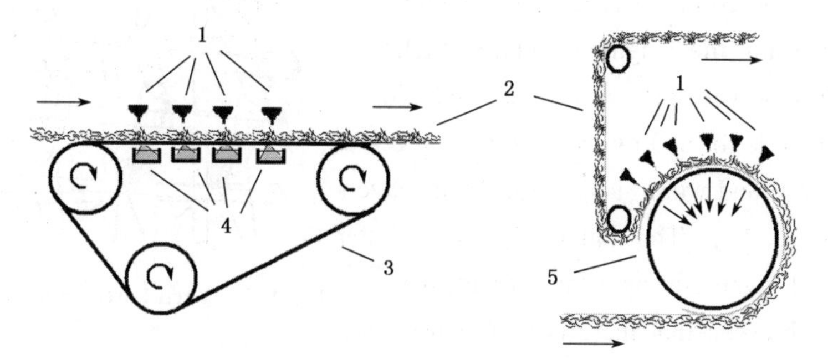

Fig. 5. 5 Illustration of water punching

1—water injectors 2—web 3—delivery apron
4—suction boxes 5—rotary screen drum with suction function

1.1.3 Stitch-bonded

Fibre webs can also be bonded with stitches. Fig. 5. 6 is an example of stitch-bonding machines. They work on the same principle as a warp knitting machine. The knitting needles, usually compound needles and sliders, penetrate the web and interact with sinkers and guides carrying yarns to produce single guide bar structures to bond the web. The stitch-bonded fabrics resemble needled felts to some extent but the stitch structure allows adequate strength and directional control to be achieved without undue stiffness.

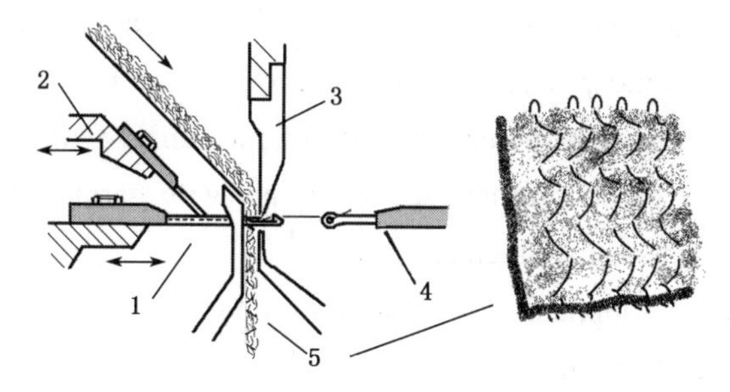

Fig. 5. 6 Illustration of stitch bonding

1—compound needle 2—slider 3—sinker 4—guide 5—stitch bonded web

1.2 YARN-BASED NON-WOVEN FABRICS

In yarn-based non-woven fabrics, layers of yarns instead of fibre webs are used as the substrates. One common type of these fabrics is formed by using the lengthwise chain stitches to bond widthwise yarn-layers together, and another common type is formed by using tricot stitches to reinforce weft and warp bi-directional yarn-layers (see Fig. 5. 7). The latter type has better stability and higher strength than the former.

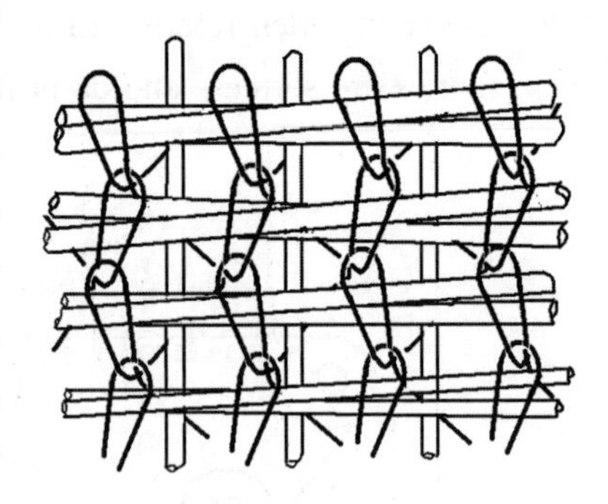

Fig. 5.7 Yarn-based non-woven fabric

It can be seen that there is in fact no clear-cut difference between this type of non-wovens and some sort of laid-in warp knitted structures like multi-axial warp knitted fabrics, and nowadays some fibre hybrid composites are made with multi-axial warp knitting technology to achieve very good tensile and impact properties.

2 APPLICATION OF NON-WOVEN FABRICS

It can be seen that, compared with the production for knitted or woven fabrics, the production sequence for non-woven fabrics can be much shorter and the production output can also be much greater. With the development of the technology, the applications for non-woven fabrics are becoming wider and wider.

One of their traditional applications is in making garment accessories such as filling, wadding or shoulder padding, and base cloths for interlinings. The fact that they can be made voluminous to entrap a lot of air can give them heat preservation property and that they can be cut and sewn together with garment linings makes them popular in winter garment manufacture.

Another traditional application of non-woven fabrics is to use them as the base cloths for artificial leathers and carpets. This can reduce the production cost whilst at the same time produce a similar quality to those using a woven base cloth. A similar application can be found in upholstery or automobile interiors such as mattresses, seat cushions, wall paper and linings.

Due to their porosity, non-woven materials have been widely used as filter materials. Non-woven filters are now widely used, not only in domestic and industrial fields but also in various medical or surgery support systems because of the tremendous developments in fibre science, web forming and the fabric forming technologies for non-woven fabrics. For the same reason and because of the relatively lower cost, non-woven fabrics are also widely used as geotextiles in the foundation constructions for buildings, roads, railways and river banks.

Due to its very good absorbency, non-woven material is widely used in making hygienic goods like wipes, such as dry or wet napkins, and diapers for babies and incontinent adults. In addition to the consideration of absorbency and strength, nonwovens producers have been working on developing wipes flushable and biodegradable.

Through suitable finishing processes the mechanical properties of some hydroentangled fabrics can be made as good as some woven or knitted fabrics in terms of strength, porosity, drape and laundering; therefore some can be used as apparel fabrics and thus reduce the garment cost.

In the field of medicine, non-woven fabrics are also used in making disposable facemasks and operating gowns for doctors and nurses. In addition, non-woven medical dressings and bandages are also becoming popular in hospitals. Impregnated with antimicrobial agents, these can greatly reduce infections.

Furthermore, the application of non-woven fabrics has been extended to the military and hi-technology fields and they are now used as the base materials for shell casings in planes and space shuttles, and bullet-proof clothing as well.

The market for non-woven fabrics continues to grow as more development and improvement of the non-woven fabrics take place and, consequently, it is likely that their applications will become even more widespread.

Words and Phrases

web-based non-woven fabric 以纤维网为底的非织造织物

carding	梳理成网
wetlaying	湿法成网
airlaying	气流成网
fibre web	纤维网
stability	稳定性
composite web	复合纤维网
bonded non-woven fabric	黏合法形成的非织造织物
acrylic [əˈkrɪlɪk] bonding-agent	聚丙烯黏合剂
acrylate [ˈækrɪleɪt] bonding-agent	聚丙烯酸酯黏合剂
immerse [ɪˈmɜːs]	浸渍
heat cylinder	热滚筒
hot airflow oven	热风烘炉
porosity [pɔːˈrɒsɪtɪ]	透气性,多孔性
handle	手感
wall covering	墙布
disposable [dɪsˈpəʊzəbl] product	一次性(使用)产品
endless delivery apron [ˈeɪprən]	循环的输网帘
rotary screen drum	圆网滚筒
cleaning trough [ˈtrɔːf]	清洗槽
immersion trough	浸渍槽
squeezing roller	轧辊
heat bonding	热熔黏合
thermo-fusible fibre	热熔性纤维
engrave [ɪnˈgreɪv]	雕刻
emboss [ɪmˈbɒs]	凹凸纹
bulky [ˈbʌlkɪ]	蓬松的
filtration [fɪlˈtreɪʃən]	过滤
wadding [ˈwɒdɪŋ]	填料
quilt [kwɪlt]	被子
filter cloths	过滤布
tufted [ˈtʌftɪd] rug	簇绒地毯
spun-bonded fabric	纺黏非织造织物

spinneret [ˈspɪnəret] 喷丝头,喷丝板
laminate [ˈlæmɪneɪt] 薄层
needle punching 针刺
carpet backing 地毯底布
drafting device 牵伸装置
condenser screen 凝网帘
fibre entangling [ɪnˈtæŋglɪŋ] type 纤维缠结型
hydroentanglement [ˈhaɪdrəʊɪnˈtæŋglmənt] 射流缠结(水刺)
barbed [bɑːbd] needle 刺针,倒钩针
needle felting [ˈfeltɪŋ] 针刺
dry-laid web 干法成网
wet-laid web 湿法成网
geotextile [ˌdʒiːəʊˈtekstaɪl] 土工布
base cloth 底布
artificial leather 人造革
carpet tile 拼接式地毯
spun-lace fabric 水刺非织造织物
compression roller 压网辊
compression plate 压网板
web plate 托网板
needled fabric 针刺织物
water injector 水针
stitch-bonded fabric 缝编织物
stitch-bonding machine 缝编机
yarn-based non-woven fabric 以纱线为底的非织造物
substrate [ˈsʌbstreɪt] 底布,衬底
fibre hybrid composites 纤维混合型复合织物
production sequence [ˈsiːkwəns] 生产流程
production output 产量
garment accessory [əkˈsesərɪ] 服装辅料
garment filling 服装芯料
shoulder padding 肩衬

interlining [ˈɪntəˈlaɪnɪŋ] 衬头
heat preservation [ˌprezə(ː)ˈveɪʃən] property 保暖性
garment lining 服装里料
automobile interiors [ɪnˈtɪərɪə] 车内装潢
mattress [ˈmætrɪs] 垫子，踏脚垫
seat cushion [ˈkʊʃən] 座垫
wall paper 墙纸
wall lining 墙衬
surgery [ˈsɜːdʒərɪ] support system 手术支持系统
foundation [faʊnˈdeɪʃən] construction 基础建设
absorbency [əbˈsɔːbənsɪ] 吸附性
hygienic [haɪˈdʒiːnɪk] goods 卫生用品
wipes 擦拭巾，擦拭纸
napkin [ˈnæpkɪn] 餐巾，纸巾
diaper [ˈdaɪəpə] 尿布
incontinent [ɪnˈkɒntɪnənt] adult 大小便失禁的成人
flushable 能冲散的
biodegradable 能生物降解的
facemask 口罩
operating gown 手术衣
medical dressings [ˈdresɪŋz] 医用敷料
bandage [ˈbændɪdʒ] 绷带
antimicrobial [ˌæntɪmaɪˈkrəʊbɪəl] agent 抗菌剂
infection [ɪnˈfekʃən] 感染
shell casing 壳体

Exercises

1. By which of the following ways can bonded non-woven fabrics be produced? (　　)

   a) By applying bonding agent　　b) By needle punching

c) By stitch-bonding d) By applying heat and pressure

2. Which of the following fibres is suitable for making non-woven fabrics through heat bonding? ( )
   a) Wool b) Rayon
   c) Polyamide d) Cotton
3. For which type of non-woven fabrics, may a warping process be necessary? ( )
   a) Needled non-woven fabric
   b) Spun-bonded non-woven fabric
   c) Thermally bonded non-woven fabric
   d) Stitch bonded non-woven fabric
4. Which of the following non-woven fabrics have good directional tensile strength? ( )
   a) Needle punched non-woven fabric
   b) Thermally bonded non-woven
   c) Stitch bonded non-woven fabric
   d) Yarn-based non-woven fabric
5. To produce non-woven fabrics, fibrous webs could usually be prepared through the process of ________. ( )
   a) carding b) combing c) wetlaying d) airlaying

Reading Materials

Comprehensive Spunbond Portfolio — always the right solution

For industrial nonwovens, Oerlikon Nonwoven systems are capable of high production capacities and yields with simultaneously low energy consumption. To this end, geotextiles made from polypropylene or polyester can be efficiently manufactured with running meter weights of up to 400 g/m^2 and filament titers of up to 9 dtex, for example. And Oerlikon Nonwoven also offers specialized spunbond processes for producing nonwoven substrates for roofing underlay (PP or PET spunbonds) and so-called bitumen roofing substrates (needled PET spunbonds) for bitumen roofing membranes. Furthermore, spunbond products are also becoming increasingly important in filtration applications both as backing materials for filter media and as the filter media themselves. A flexible nonwoven structure permits the inclusion of customer-specific requirements for various functions. It is Oerlikon Nonwoven's many years of core-sheath bi-component experience in particular that enable the creation of completely new nonwoven structures and hence the incorporation of various functions in a single material. The core-sheath bi-component spinning process permits various combined fiber cross-sections and also simultaneously different fibers to be produced from a single or different polymers. The spectrum ranges from core-sheath and side-by-side bi-component filaments, splitable fibers all the way through to so-called mixed fibers.

— excerpted from *Technical Textiles Innovation*, October-December 2021, page 14

【参考提示】

1. *Technical Textile Innovation* 是印度 Times International 公司出版的电子季刊。
2. spunbond portfolio,纺黏系列产品。
3. Oerlikon,总部在瑞士的世界知名工业集团,国内有译为"欧瑞康"的。Oerlikon Nonwoven 是该集团旗下的一个重要品牌。
4. ... running meter weights of up to 400 g/m^2,原文如此。可能作者是想说"... running square meter weights..."吧。通常的表达方式应该是"... running weights per square meter..."。
5. titer,纤度。
6. roofing underlay,屋面衬垫材料;bitumen,沥青。
7. core-sheath bi-component,芯鞘型双组分;side-by-side bi-component,并列型双组分。

Nonwovens for the Automotive Market

Tara Olivo, Associate Editor

Opportunities for nonwovens in the automotive market continue to expand. As auto designers look for materials that are lightweight, cost-effective, sustainable and offer increased comfort and reduced noise to drivers and passengers, nonwovens are helping meet these needs. Add to this the anticipated growth in the electric vehicle

(EV) market, and suppliers and converters of nonwovens are optimistic about this market.

"The role of nonwovens is indeed expanding due to their versatility and ability to be adapted to new applications," says Gerhard Klier, sales director, Technical Products for Sandler, which manufactures a range of nonwovens for automotive applications. "Most importantly, in most of the cases nonwovens are a sustainable choice."

Nonwovens can combine different properties in a single material, such as a good sound absorber can be a good thermal insulator as well, he says. Nonwovens can also be multifunctional, like flat sheets as, for example, sound absorbing pads, but also as a two-dimensional die cut part or as a three-dimensional molded part. "As such, single-polymer materials can be used, offering combinations of characteristics otherwise only achievable through combinations of different products. Made from raw materials such as polyester, they are easy-to-handle, versatile, and suitable for recycling and reuse. A broad range of technologies for producing such textiles expand the areas of application and offer more choice for designers and construction engineers."

— excerpted from *Nonwovens Industry*, December 2021, page 30

【参考提示】

1. *Nonwovens Industry* 是美国 Rodman Media 公司出版的介绍非织造物生产、市场及专利方面的单月刊杂志。
2. Sandler,德国知名的非织造产品生产商。国内有译为“盛德”的。
3. converters of nonwovens,非织造物制品生产商(以非织造物为原料的产品制造商)。
4. two-dimensional die cut part,二维的模具冲切部件。
5. three-dimensional molded part,三维的模压部件。

CHAPTER 6

DYEING AND FINISHING

Dyeing and finishing are critical processes in the manufacture of textiles because they impart colour, appearance and handle to the final product. The processes depend on the equipment used, the constituent materials and the structure of the yarns and fabrics. Dyeing and finishing may be carried out at various stages in textile production.

Natural fibres such as cotton or wool may be dyed before being spun into yarns and yarns produced in this way are called fibre-dyed yarns. Dyes could be added to the spinning solutions or even in the polymer chips when man-made fibres are spun, and, in this way, solution-dyed yarns or spun-dyed yarns are made. For yarn-dyed fabrics, yarns need to be dyed before weaving or knitting takes place. Dyeing machines are designed for dyeing yarns in the form of either loosely wound hanks or wound onto packages. Such machines are referred to as hank dyeing and package dyeing machines respectively.

Finishing processes may also be performed on the assembled garments. For example, denim clothing washed in many ways, such as stone washing or enzyme washing, is very popular these days. Garment dyeing might also be used for some types of knitwear to produce garments so as to avoid colour shading within them.

However, in most cases dyeing and finishing are carried out on fabrics, whereby cloths are woven or knitted and then these grey or "greige" state fabrics, after preliminary treatments, are dyed, and/or printed, and chemically or mechanically finished.

1 PRELIMINARY TREATMENTS

In order to achieve "predictable and reproducible" results in dyeing and finishing, some preliminary treatments are necessary. Depending on the process, fabrics may be treated as single pieces or batches, or sewn together using chain stitches, easily to be removed for post-processing, to create long lengths of different batches for continuous processing.

1.1 SINGEING

Singeing is the process to burn off fibres or nap on the fabric surface to avoid uneven dyeing or printing blotches. Generally speaking, woven cotton grey cloths need to be singed before other preliminary treatments are started. There are several types of singeing machines, such as the plate singer, the roller singer and the gas singer. The plate singeing machine is the simplest and oldest type. The cloth to be singed passes over one or two heated copper plates at high speed to remove the nap but without scorching the cloth. In the roller singeing machine, heated steel rollers are used instead of the copper plates to give better control of the heating. The gas singeing machine, in which the fabric passes over gas burners to singe the surface fibres, is the most commonly used type nowadays. The number and position of the burners and the length of the flames can be adjusted to achieve the best result.

1.2 DESIZING

As mentioned in Chapter 3, for warp yarns, especially cotton, used in weaving, sizing, usually using starch, is generally necessary to reduce the yarn hairiness and strengthen the yarn so that it can withstand the weaving tensions. However the size left on the cloth may hinder the chemicals or dyes from contacting the fibres of the cloth. Consequently the size must be removed before scouring starts.

The process to remove the size from the cloth is called desizing or steeping. Enzyme desizing, alkali desizing or acid desizing may be used. In enzyme desizing, the cloths are padded with hot water to swell the starch, and then padded in enzyme liquor. After being stacked in piles for 2 to 4 hours, the cloths are washed in hot water. Enzyme desizing requires less time and causes less damage to the cloths, but if chemical size instead of wheat starch is used, enzymes may not remove the size. Then, the widely used method for desizing is alkali desizing. The fabrics are impregnated with a weak solution of caustic soda and piled into a steeping bin for 2 to 12 hours, and then washed. If after that, the cloths are treated with dilute sulphuric acid, better results can be achieved.

For knitted fabrics, desizing is not needed since yarns used in knitting are not sized.

1.3 SCOURING

For the grey goods made of natural fibres, impurities on the fibres are inevitable. Taking cotton as an example, there could be waxes, pectin products as well as vegetable and mineral substances in them. These impurities may give the raw fibres a yellowish colour and make them harsh to handle. The waxy impurities in the fibres and oil spots on fabrics are likely to affect the dyeing results.

Furthermore, waxing or oiling might be necessary to make the staple yarns soft and smooth with lower frictional coefficients for winding or knitting. For synthetic filaments, especially those to be used in warp knitting, surface active agents and static inhibitors, which are usually a specially formulated oil emulsion, should be used during warping; otherwise the filaments may carry electrostatic charges, which will severely disturb the knitting or weaving actions.

All impurities including oils and waxes must be removed before dyeing and finishing, and scouring can, to a great extent, serve this purpose. One of the most common methods of scouring for cotton grey cloth is kier boiling. The cotton cloth is packed evenly in a tightly sealed kier and boiling alkaline liquors are circulated in the kier under pressure. Another commonly used way in scouring is continuous steaming and the scouring is processed in serially arranged apparatus, which generally comprises a mangle, a J-box and a roller washing machine.

It can be seen from Fig. 6. 1 that the alkaline liquor is applied onto the fabric through the mangle, and then, the fabric is fed into the J-box, in which saturated steam is injected through the steam heater, and afterwards, the fabric is piled uniformly. After one or more hours, the fabric is delivered to the roller washing machine.

1.4 BLEACHING

Although most of the impurities in cotton or linen cloths can be removed after scouring, the natural colour still remains in the cloth. For such cloths to be dyed to a

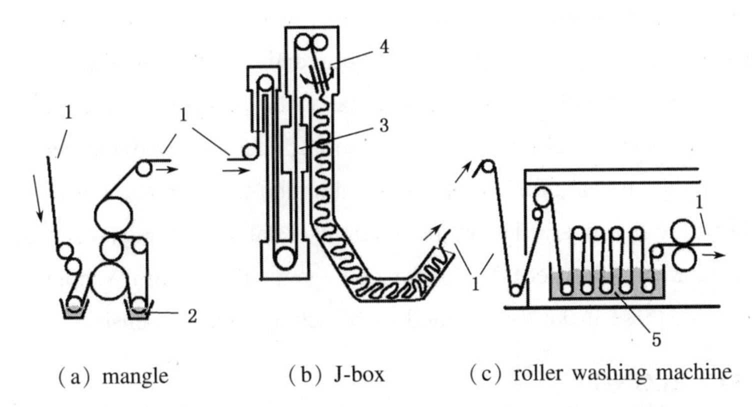

Fig. 6.1 Equipment used in continuously steaming and scouring process

1—fabric 2—trough with alkaline liquors 3—steam heater
4—cuttling or plaiting device 5—washing trough

light colour or to be used as the ground cloths for prints, bleaching is necessary to remove the inherent colour.

The bleaching agent is actually an oxidizing agent. The following bleaching agents are commonly used:

Sodium hypochlorite (calcium hypochlorite may also be used) may be the commonly used bleaching agent. Bleaching with sodium hypochlorite is generally performed under alkaline conditions, because under neutral or acidic conditions the sodium hypochlorite will be severely decomposed and the oxidization of the cellulosic fibres will be intensified, which may make the cellulosic fibres become oxidized cellulose. Furthermore, metals such as iron, nickel and copper and their compounds are very good catalytic agents in the decomposition of sodium hypochlorite; therefore equipment made of such materials cannot be used in the process.

Hydrogen peroxide is an excellent bleaching agent. There are many advantages for bleaching with hydrogen peroxide. For example, the bleached fabric will have a good whiteness and a stable structure, and reduction in fabric strength is less than that when bleached with sodium hypochlorite. It is possible to combine the desizing, scouring and bleaching processes into one process. Bleaching with hydrogen peroxide is generally performed in a weak alkali solution, and stabilizers such as sodium silicate or tri-ethanolamine should be used to overcome the catalytic actions caused by the metals mentioned above and their compounds.

Sodium chlorite is another bleaching agent, which can impart a good whiteness into the fabric with less damage to the fibre and is also suitable for continuous processing. Bleaching with sodium chlorite has to be performed in acidic conditions. However as the sodium chlorite is decomposed, chlorine dioxide vapour will be released, and this is harmful to human health and is strongly corrosive to many metals, plastics and rubber. Therefore titanium metal is generally used to make the bleaching equipment, and necessary protection against the harmful vapours would have to be taken. All these make this method of bleaching more expensive.

1.5 MERCERIZING

Mercerization is an alkali treatment applied to cotton or linen fabrics under tensions. Mercerization swells the lumen in cotton fibres to make them more lustrous to give them a silk like appearance and to enable them to accept dyes more evenly.

Tensions must be exerted on the cloth during mercerization and this is done on a stenter. The cloth is first padded with caustic soda after passing through two mangles, and then clips, attached to chains of the stenter, grip the selvedges of the cloth and move forward as the chains move. These two chains are arranged in such a way that the distance between them increases as the chains move forward. Thus, the cloth contraction due to the effect of the caustic soda and the diverging arrangement of the clip chains build up the gripping tension to stretch the cloth. The speed of these two chains can be adjusted separately to align the warp and weft correctly. Before leaving the stenter, the cloth will be washed by spraying water onto the fabric and then vacuum extracting the mercerizing liquor and the washing liquor, so that, when it leaves the stenter, the cloth will not have a tendency to contract.

Mercerization could be performed before bleaching in order to enhance the whiteness for fabrics such as bleached cotton poplin and cotton prints. When dyeing heavy cloths, mercerization after bleaching is preferred to prevent crease marks in the dyed cloth.

1.6 HEAT SETTING

Heat setting might be necessary for most fabrics made of synthetic fibres or their blends. In heat setting, the fabrics will be stretched or compacted to a desired dimension and then heat set. Heat setting can remove the crease marks on the fabric and improve fabric heat stability, mechanical strength and hand-touch, and besides, it will improve the fabric dyeing property.

Both wet heat setting and dry heat setting are used. In wet heat setting, water is used as a swelling agent to improve the setting effect; therefore fabrics made from either 100% or blends of polyamide may be wet heat setting. However fabrics made from 100% or blends of polyester are dry heat set because of the hydrophobicity of polyester.

The dry heat setting machine is an open width stenter with a chain feed for the fabric. When heat setting, the cloth is usually overfed onto the pin clips, which hold the selvedges of the cloth. The pin clips are mounted on two chains, moving forward towards the hot air chamber. The distance between the two chains is adjustable for the desired fabric width and the speed of the chains can also be adjusted to control the heat setting time. The heat setting temperatures of around 170℃ to 210℃ are best suited for fabric shape retention and crease removal. As it leaves the hot air chamber, the cloth should pass through a cooling zone, where the surface temperature of the fabric will be decreased to 50℃ or below. Appropriate fabric overfeeding and widthwise tension are important as these affect the fabric heat stability and mechanical strength.

Heat setting could also be the final process in the dyeing and finishing of synthetic fabrics.

2 DYEING OF FABRICS

The purpose of dyeing a fabric is to impart evenly a solid colour to the fabric with good colour fastness. In order to achieve the best dyeing effect at a reasonable cost, the types of dye, auxiliary agents, dyeing machine availability, processing sequence and technical parameters, such as processing time and temperature, fabric weight to liquor ratio, concentration of the dyeing liquors, etc. should be carefully chosen.

2.1 DYES COMMONLY USED

For some fibres such as cellulose fibres and protein fibres, many types of dyes can be used. However for synthetic fibres and fabrics made from them, the choice of dyes is limited. The commonly used dyes for textile applications are as follows:

1) **Direct Dye** Direct dye is a hydrophilic anionic dye, which has substantivity for cellulosic fibres and protein fibres through hydrogen bonds and the Van der Waals forces. When dyeing fabrics with the direct dyes, soft water must be used, or the calcium ions or magnesium ions in the water precipitate the dyestuff. Sodium chloride could be used as an accelerant when direct dye is used for cotton fabrics. For viscose fabrics, a leveling agent is required. Cationic fixatives could be used to improve the colour fastness.

2) **Reactive Dye** Reactive dyes are also hydrophilic dyes. The active groups of the reactive dyes can bond well with the hydroxyl groups in the cellulose fibres, or with the amino groups in the protein fibres in the form of covalent bonds. Fabrics dyed with reactive dyes provide good brightness and good colour fastness to laundering and rubbing.

3) **Vat Dye** Vat dyes are available as soluble and insoluble types. If the latter is used, it should be first made soluble through the reduction of its carbonyl groups in an alkaline aqueous solution to a leuco-compound with hydroxyl groups, which is then applied to the fibres and oxidized back to its original insoluble form. The soluble vat dye is in fact a leuco dye in the scope of sodium salts with sulphuric ester. It can be dissolved in water before dyeing the cellulose fibres or protein fibres, and after being applied onto the cloth, it will be oxidized to an insoluble form in an acid bath, and then be fixed in the fibres.

4) **Sulphur Dye** Sulphur dye is a dye derived from chemicals containing sulphur. It is used mostly for cellulose fibres. It has a fair resistance to washing but poor resistance to sunlight. It is one of the cheapest dyes but the sulphidation might weaken cellulose fibres.

5) **Acid Dye** Acid dye is an anionic dye, most of which is aromatic sodium salts of sulphonic acids. It is used on protein fibre, such as wool and silk, and on polyamide and modified acrylics such as Orlon 44.

6) **Disperse Dye** Disperse dye is a hydrophobic and non-ionic dye. It was developed with the development of man-made fibres. The early generations of disperse dye were developed for cellulose acetate, and now disperse dyes are widely used in dyeing hydrophobic synthetic fibres like polyester. The dyestuff is taken up due to its

very simple and small molecular structure. It has substantivity for hydrophobic fibres and its application requires high pressures and temperatures.

7) **Cationic Dye** Cationic dye is also called a basic dye and it can be dissociated in an aqueous solution to produce a positively charged colour ion. It is used to dye acrylic fibres and certain modified polyester fibres such as Dacron 64.

In addition to the dyes, many other auxiliary agents have to be added as appropriate to achieve good dyeing results. Commonly used auxiliary agents are:

- wetting agents which are used to promote rapid penetration of the processing liquors into the fibres;
- softeners which are used to overcome the harshness of the fabrics and imparted during dyeing or other treatments;
- dispersing agents which can break down the dye particles into smaller ones to facilitate their penetration into the fibres;
- dye-carriers which swell hydrophobic fibres and make them more porous to take up dyes;
- sequestrants which combine the metal impurities and make them inactive against the dyes being used;
- dye-fixing agents which make dyes colour fast in the fabric.

Furthermore, detergents, leveling agents, anti-static agents, anti-foaming agents, rot-proofing agents, moth-proofing agents, emulsifiers and mordants, etc. may also be used. Some auxiliary agents are also used in the printing or finishing process. The agent used depends on the type of fabric to be dyed, the type of dye used and the type of chemical reactions involved.

2.2 FABRIC DYEING MACHINES

Dyeing machines are selected mainly according to the type of fabrics to be dyed and the required dyeing effect.

1) **Jig Dyeing Machine** Jig dyeing is suitable for fabrics made from cellulosic fibres. The jig dyeing machine has a trough containing dye liquor. Above the trough there is a pair of batching rollers, which can rotate in clockwise or anti-clockwise directions alternately, and at the bottom of the trough there are guide rollers. Fabrics to be dyed are wound onto one of the batching rollers and unwound onto the other,

and vice versa. The guide rollers ensure the cloths are immersed in the dye bath and steam pipes arranged at the bottom of the trough can heat the dye liquor to the desired temperature (see Fig. 6.2).

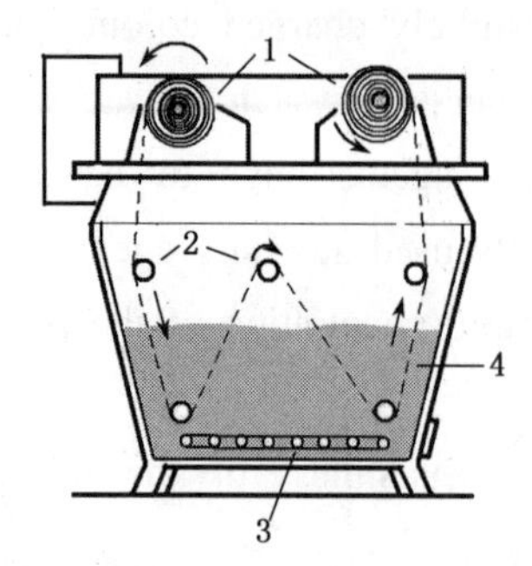

Fig. 6.2 Jig dyeing machine
1—fabric batching rollers 2—guide rollers
3—steam pipes 4—dye liquor

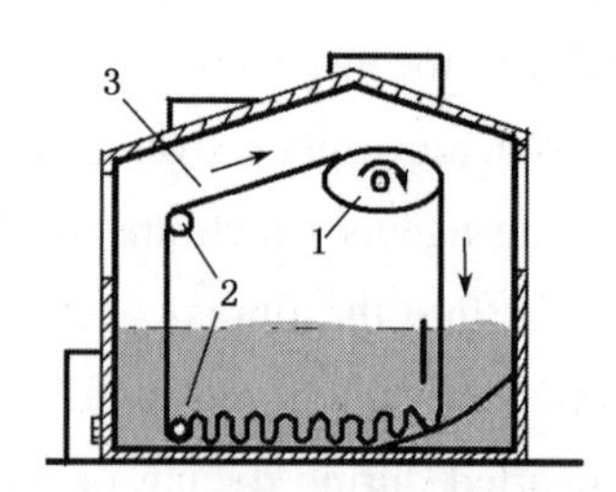

Fig. 6.3 Winch dyeing machine
1—winch 2—guide rollers 3—fabric

2) **Winch Dyeing Machine** If the fabric must not be subjected to any lengthwise tension, a winch dyeing machine can be used. The dye vat of the machine is filled with dye liquor and fabrics being dyed are submerged in the dye liquor. The rotation of the winch keeps the fabrics moving in a tensionless state (Fig. 6.3).

3) **Pad Dyer** Pad dyeing is one of the most popular dyeing methodologies, in which cloths can be processed continuously. There are different types of pad-roll systems and generally, the pad-roll system consists of a two or three-roller padder, an infrared heating channel and several reaction chambers. The cloth is impregnated in the padder and heated by the infrared radiators, and then enters the reaction chambers. The temperature in the chambers is maintained at a desired level through the injection of steam. After the reaction chamber is a series of open-width washers where soaping and rinsing take place. Since pad dyeing will cause greater tensions on the cloth being processed, it is not suitable for highly extensible fabrics such as knitted fabrics and elastomeric fabrics.

4) **High Temperature-pressure Dyeing Machines** Generally speaking, fabrics made from synthetic fibres such as polyester should be dyed under high temperature and high pressure so that the dye can quickly and efficiently penetrate into the fibres. There are several types of such machines, for example, the high temperature and

pressure winch beck, the high temperature and pressure overflow dyeing machine and the high temperature and pressure jet dyeing machine.

In the high temperature and pressure overflow dyeing machine, the cloth being dyed is pushed forward by the flow of the dyeing liquor; therefore little tension would be exerted on the cloth and better dyeing results could be obtained (see Fig. 6.4). In the high temperature and pressure jet dyeing machine, the function of the jet device enables the dyes to penetrate into the fibres. Based on the working principles of the above two, the high temperature and pressure overflow jet dyeing machine was designed which is very useful for dyeing elastomeric synthetic fabrics.

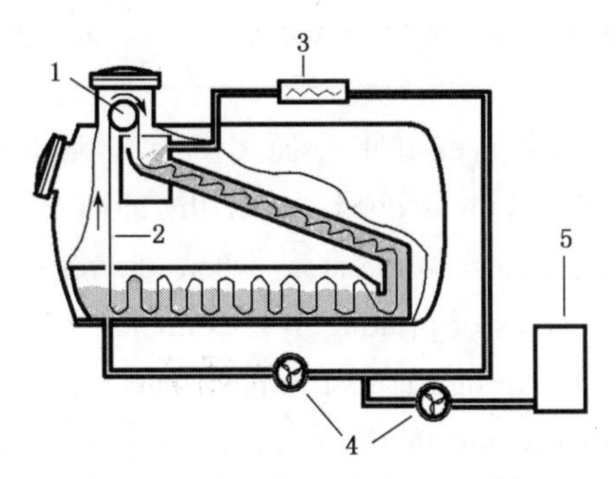

Fig. 6.4 Overflow dyeing machine

1—guide roller 2—fabric 3—heater 4—pumps 5—replenisher for dyeing liquor

5) **Other Equipment** Other equipment is necessary to complete a dyeing process, for example, an open-width washer or rope washer to wash away the loose colour or perform the neutralization; a centrifugal hydroextractor or vacuum hydroextractor to remove excessive water and a rotary screen dryer or short loop hanging dryer to dry the cloth, etc.

3 PRINTING

3.1 PRINTING METHODS

Technologically, there are several methods of printing, such as direct printing, discharge printing and resist printing.

In direct printing, printing paste should first be prepared. Pastes, such as alginate paste or starch paste, need to be mixed in the required proportions with dyes and the other necessary chemicals such as wetting agents and fixing agents. These are then printed on white ground cloth according to the desired designs. For synthetic fabrics, the printing paste could be made with pigments instead of dyes, and then the printing paste would comprise of pigments, adhesives, emulsion paste and other necessary chemicals.

In discharge printing, the ground cloth should first be dyed with the desired ground colour, and then the ground colour is discharged or bleached in different areas by printing it with the discharge paste to leave the desired white designs. The discharge paste is usually made with reducing agent such as sodium sulphoxylate-formaldehyde.

In resist printing, substances that resist dyeing should first be applied on the ground cloth, and then the cloth is dyed. After the cloth is dyed, the resist will be removed, and the designs appear in the areas where the resist was printed.

There are also other types of printing, for example, sublistatic printing and flock printing. In the former, the design is first printed onto paper and then the paper with the designs is pressed against the fabric or garments such as T-shirts. When heat is applied, the designs are transferred onto the fabric or garment. In the latter, short fibrous materials are printed in patterns onto fabrics with the help of adhesives. Electrostatic flocking is commonly used.

3.2 PRINTING EQUIPMENT

Printing may be performed by roller printing, screen printing or, more recently, inkjet printing equipment.

3.2.1 Roller Printing

A roller printing machine typically comprises a large central pressure cylinder (or called as pressure bowl) covered with rubber or several plies of wool-linen blended cloth which provide the cylinder with a smooth and compressively elastic surface. Several copper rollers engraved with the designs to be printed are set around the pressure cylinder, one roller for each colour, in contact with the pressure cylinder. As they rotate, each engraved printing roller, driven positively, also drives its furnisher roller, and the latter carries the printing paste from its colour box to the engraved printing roller. A sharp steel blade called a cleaning doctor blade removes the excess paste from the printing roller, and another blade called a lint doctor blade scrapes off any lint or dirt caught by the printing roller. The cloth to be printed is fed between the printing rollers and the pressure cylinder, together with a grey backing cloth to prevent the surface of the cylinder from being stained if the colouring paste penetrates the cloth (see Fig. 6. 5).

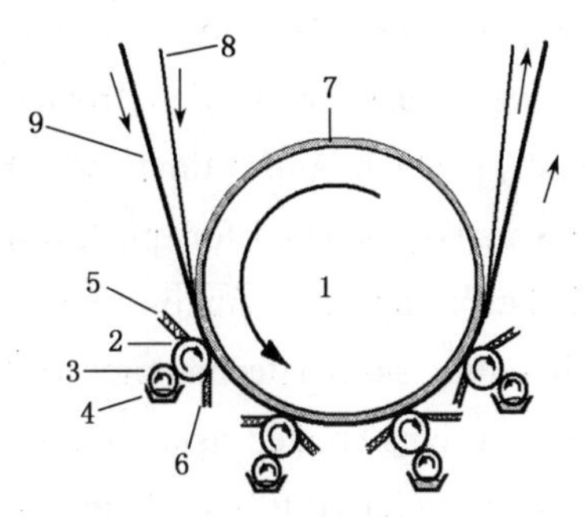

Fig. 6. 5 Roller printing

1—pressure bowl 2—printing roller 3—furnisher roller 4—colour box
5—cleaning doctor blade 6—lint doctor blade 7—rubber cover 8—backing cloth 9—fabric

Roller printing can offer a very high productivity but the preparation of the engraved printing rollers is expensive, which, practically, makes it only suited to long production runs. Furthermore, the diameter of the printing roller limits the pattern size.

3.2.2 Screen Printing

Screen printing, on the other hand, is suitable for smaller orders, and is particularly suitable for printing stretch fabrics. In screen printing, the woven mesh printing screens should first be prepared according to the designs to be printed, one for each colour. On the screen, areas where no colouring paste should penetrate are coated with insoluble film leaving the remaining screen interstices open to allow print paste to penetrate through them. Printing is done by forcing the appropriate printing paste through the mesh pattern onto the fabric underneath. The screen is prepared by coating the screen with photogelatin first and superimposing a negative image of the design onto it and then exposing it to light which fixes an insoluble film coating on the screen. The coating is washed off from those areas where the coating has not been cured, leaving the interstices in the screen open. Traditional screen printing is flat screen printing, but rotary screen printing is also very popular for larger productivity.

3.2.3 Inkjet Printing

It can be seen that for either roller printing or screen-printing the preparation is time and money consuming even though Computer Aided Design (CAD) systems have widely been used in many printing factories to assist in the design preparation. Designs to be printed must be analyzed to decide what colours could be involved, and then negative patterns are prepared for each colour and transferred to printing rollers or screens. During screen printing in mass production, rotary or flat, screens need to be changed and cleaned frequently, which is also time and labour consuming.

In order to meet today's market demand for quick response and small batch sizes inkjet printing technology is becoming increasingly used.

Inkjet printing on textiles uses similar technology to that used in paper printing. The digital information of the designs created using a CAD system can be sent to the inkjet printer (or more commonly referred to as a digital inkjet printer, and the textiles printed with it may be called as digital textiles) directly and printed onto the fabrics. Compared with the traditional printing technologies, the process is simple and less time and skill are required as the process is automatic. Furthermore, less pollution will be caused.

Generally speaking, there are two basic principles for inkjet printing for textiles. One is the Continuous Ink Jetting (CIJ) and the other is called "Drop on Demand" (DOD). In the former case, a very high pressure (around 300 kPa) built up through the ink supply pump forces the ink continuously to the nozzle, the diameter of which is usually about 10 to 100 micrometres. Under high frequency vibration caused by a piezoelectric vibrator, the ink is then broken into a flow of droplets and ejected from the nozzle at very high speed. According to the designs, a computer will send signals to the charge electrode which electrically charges selected ink droplets. When passing through the deflection electrodes, uncharged droplets will go straight into a collecting gutter whereas charged ink droplets will be deflected onto the fabric to form a part of the printed pattern (see Fig. 6.6 (a)).

In the "drop on demand" technique, ink droplets are supplied as they are needed. This can be done through an electromechanical transfer method (see Fig. 6.6 (b)). According to the patterns to be printed, a computer sends pulsed signals to the piezoelectric device which in turn deforms and produces pressure on the ink chamber through a flexible intermediary material. The pressure causes the ink droplets to be ejected from the nozzle. Another way commonly used in the DOD technique is through the electric thermal transfer method (see Fig. 6. 6 (c)). In response to computer signals the heater generates bubbles in the ink chamber, and the expansive force of the bubbles causes ink droplets to be ejected.

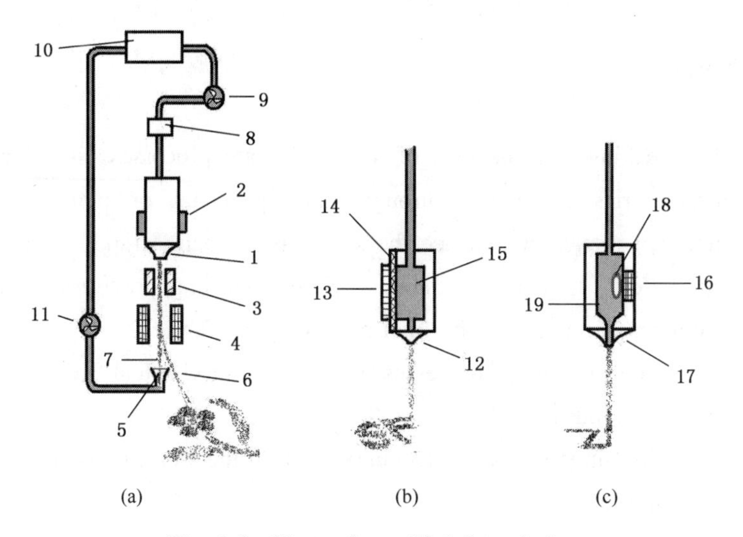

Fig. 6. 6 Illustrations of ink jet printing

(a) CIJ printing (b) DOD printing (electromechanical transfer method)
(c) DOD printing (electric thermal transfer method)
1—nozzle 2—piezoelectric vibrator 3—charge electrode 4—deflection electrode
5—collecting gutter 6—charged ink droplets 7—uncharged ink droplets 8—pressure regulator
9—ink supply pump 10—ink bottle 11—recycling pump 12—nozzle 13—piezoelectric device
14—flexible material 15—ink chamber 16—heater 17—nozzle 18—bubble 19—ink chamber

The DOD technique is cheaper but the printing speed is also lower than that of the CIJ technique. Since the ink droplets are ejected continuously, nozzle clogging problems will not occur under the CIJ technique.

Inkjet printers usually use a combination of four colours, that is, **c**yan, **m**agenta, **y**ellow and blac**k** (CMYK), to print designs with various colours, and therefore four printing heads should be assembled, one for each colour. However some printers are equipped with 2 × 8 printing heads so that theoretically up to 16 colours of ink can be printed. The print resolution of inkjet printers can reach 720 × 720 dpi. The fabrics that can be printed with inkjet printers range from natural fibres, such as cotton, silk and wool, to synthetic fibres, such as polyester and polyamide; therefore there are many types of inks needed to meet the demand. These include reactive inks, acid inks, disperse inks and even pigmented inks.

In addition to printing fabrics, inkjet printers can also be used to print T-shirt, sweatshirts, polo shirts, baby wear, aprons and towels.

4 FINISHING

Generally speaking, "finishing" of the fabrics are processes, in which desired properties of the fabrics are imparted through either mechanical or chemical methods or both. Finishing can impart or enhance the smoothness, nap, drape, lustre, gloss or crease resistance, etc. to the final fabrics. Some finishing processes such as pre-shrinkage treatments can improve the final quality of the fabric.

There are many finishing processes that apply mechanical treatments to the fabrics. Examples include:

- brushing laid-in yarns on the technical back of the weft knitted fabric to create a fleece fabric;
- napping or raising long underlaps of the warp knitted fabric to produce a pile effect;
- raising the terry loops and then cropping them to produce a velvet effect;
- calendaring fabrics over heated cylinders to impart luster;
- schreinering to create a moiré effect;
- embossing designs into fabrics.

Many woollen fabrics also need mechanical finishing such as fulling, raising and napping.

There are also many finishing processes that apply chemical treatments to fabrics, for example, crease resistant finishing, soft finishing, water repellent finishing, peaching, and finishing to give anti-static, rot-proofing or moth-proofing properties.

In some finishing processes, both mechanical and chemical treatments are involved. Coating, sand washing or stone washing are good examples.

As mentioned at the beginning of this chapter, finishing processes may also be performed on garments, and nowadays a lot of casual denim clothing undergoes sand washing, stone washing or enzyme washing after garment making-up.

Finishing can greatly improve the appearance and quality of fabrics or garments, and that is why it is one of the key processes in textile manufacture and continues to be the subject of much research.

5 QUALITY ISSUES IN DYEING AND FINISHING

Faults in finished fabrics may arise from operation malfunctions, defects in the machinery or materials and incorrect technical specifications. If the operator does not strictly follow the procedures specified in the technical schedule, or if the machine parts do not function well, or if the bath-ratio is not properly set etc, apparent faults such as listing, tailing, streakiness, coloured spots or dyeing unevenness etc. may occur. The so-called listing is a dyeing fault referred to as colour shading appearing at the selvedges of the dyed fabrics and tailing is a dyeing fault referred to as colour shading appearing at the beginning or end of the dyed piece goods. On some modern dyeing machines, dye and chemical dosing could be fulfilled automaticly according to pre-programmed times and curves. This greatly reduces the dyeing faults due to the operator's mistakes.

The inherent quality of the finished fabric should generally be tested according to standards. Colour fastness for example, is assessed visually and graded. For colour fastness to light, Grade 1 is the poorest and Grade 8 is the best; for other colour fastness, say to laundering, rubbing, weathering, or perspiration, Grade 1 is the poorest and Grade 5 is the best.

Some faults may arise from the grey cloth itself. For example, as mentioned in the previous chapter, if man-made warp yarns from different production lots are woven into a piece of cloth, vertical colour lines may appear if the dye uptake of the yarns varies from lot to lot.

In addition to colour fastness, other properties are measured in the finished fabric to check whether they are within tolerance and conform to the technical specification. These include: dimensional change, tensile strength, tear strength, yarn slippage, pill resistance, angle of crease recovery, water repellency, yarn linear density, number of warps or wales per unit width, number of wefts or courses per unit length, fabric weight per unit area, etc. The specific properties/ parameters that should be checked or tested depend on the agreement between the buyer and the supplier. It is better for both parties to clarify in the agreement the relevant standards that should be followed and what tolerances should be allowed. Attention must be paid to the fact that some aspects such as the residual formaldehyde and residual pesticide, etc. may not included in a commercial agreement, in which case, the parties concerned must strictly control them according to the government regulations.

Words and Phrases

dye	染色,染料
finishing	整理
fibre-dyed yarn	色纺纱
solution-dyed yarn	纺(前)染(色)纱,纺液着色纱
spun-dyed yarn	纺(前)染(色)纱,纺液着色纱
yarn-dyed fabric	色织布
hank	绞纱
hank dyeing machine	绞纱染色机
package dyeing machine	筒子纱染色机
stone washing	石洗
enzyme [ˈenzaɪm] washing	酵素洗
predictable [prɪˈdɪktəb(ə)l]	可预见的
reproducible [ˌriːprəˈdjuːsəbl]	可复制的
preliminary [prɪˈlɪmɪnərɪ] treatment	预处理
chain stitch	链缝缝子
singeing [sɪndʒɪŋ]	烧毛
nap	毛绒,拉毛
singeing machine	烧毛机
plate singer [ˈsɪndʒə]	平板式烧毛机
roller singer	圆筒烧毛机

gas singer	燃气烧毛机
scorching [ˈskɔːtʃɪŋ]	烤焦
gas burner [ˈbɜːnə]	火口
flame	火焰
desizing [diːˈsaɪzɪŋ]	退浆
steeping [ˈstiːpɪŋ]	浸渍,浸泡
enzyme desizing	酶退浆
alkali [ˈælkəlaɪ] desizing	碱退浆
acid desizing	酸退浆
liquor [ˈlɪkə]	液,溶液
solution [səˈljuːʃən]	溶液
caustic [ˈkɔːstɪk] soda [ˈsəʊdə]	烧碱
dilute [daɪˈljuːtˌdɪˈl-]	稀的
sulphuric acid [ˈæsɪd]	硫酸
scouring [ˈskaʊrɪŋ]	煮练(精练)
pectin [ˈpektɪn] product	果胶产物
vegetable and mineral substances	植物及矿物性物质
harsh [hɑːʃ]	粗糙
waxing	上蜡
frictional coefficient	摩擦系数
surface active agent	表面活性剂
static [ˈstætɪk] inhibitor [ɪnˈhɪbɪtə(r)]	抗静电剂
specially formulated [ˈfɔːmjʊleɪtɪd]	特别配方的
oil emulsion [ɪˈmʌlʃən]	乳化液,乳化油,油乳胶
electrostatic [ɪˈlektrəʊˈstætɪk] charge	静电荷
kier [kɪə]	煮布锅
to be circulated [ˈsɜːkjʊleɪtɪd]	被循环
continuous [kənˈtɪnjʊəs] steaming	连续汽蒸
mangle [ˈmæŋgl]	轧液机
J-box	J 型箱
roller washing machine	平洗机

cuttling [ˈkʌtlɪŋ] device	甩布装置
plaiting [plætɪŋ, pleɪtɪŋ] device	甩布装置
saturated [ˈsætʃəreɪtɪd] steam	饱和蒸汽
bleaching	漂白
oxidizing [ˈɒksɪdaɪzɪŋ] agent	氧化剂
sodium [ˈsəʊdjəm, -dɪəm] hypochlorite [ˈhaɪpəʊˈklɔːraɪt]	次氯酸钠
calcium [ˈkælsɪəm] hypochlorite	次氯酸钙
bleaching agent	漂白剂
alkaline [ˈælkəlaɪn] condition	碱性条件
neutral [ˈnjuːtrəl]	中性的
decompose [ˌdiːkəmˈpəʊz]	分解
oxidization [ˈɒksɪdaɪˈzeɪʃən; -dɪˈz-]	氧化
cellulosic [ˌseljʊˈləʊsɪk] fibre	纤维素纤维
oxidized cellulose	氧化纤维素
catalytic agent [ˈeɪdʒənt]	催化剂
hydrogen [ˈhaɪdrədʒən] peroxide [pəˈrɒksaɪd]	过氧化氢
whiteness [ˈ(h)waɪtnɪs]	白度
stabilizer [ˈsteɪbɪlaɪzə]	稳定剂
sodium silicate [ˈsɪlɪkɪt]	硅酸钠
tri-ethanolamine [ˌtraɪeθəˈnɒləmiːn]	三乙醇胺
sodium chlorite [ˈklɔːraɪt]	亚氯酸钠
acidic [əˈsɪdɪk] condition	酸性条件
chlorine [ˈklɔːriːn] dioxide [daɪˈɒksaɪd]	二氧化氯
titanium [taɪˈteɪnjəm, tɪ-]	钛
mercerization [ˌmɜːsəraɪˈzeɪʃən]	丝光
stenter [ˈstentə]	拉幅机
diverging [daɪˈvɜːdʒɪŋ] arrangement	发散式配置
mercerizing [ˈmɜːsəraɪzɪŋ] liquor	丝光液
washing liquor	洗涤液
heat setting	热定形

crease mark	皱痕
hand-touch	手感
swelling agent	溶胀剂
to be overfed	被超喂
pin clips	针排
hot air chamber	热风烘房
shape retention	保形性
overfeeding	超喂
solid colour	素色,单色
colour fastness [ˈfɑːstnɪs]	色牢度
auxiliary [ɔːgˈzɪljərɪ] agent	助剂
dyeing machine	染色机
concentration	浓度
liquor ratio	浴比
direct dye	直接染料
anionic [ˌænaɪˈɒnɪk] dye	阴离子染料
substantivity [ˌsʌbstənˈtɪvɪtɪ]	直接上染性
hydrogen bond	氢键
Van der Waals [ˈvæn dɜːˌwɔːlz] force	范德华力
soft water	软水
calcium ion [ˈaɪən]	钙离子
magnesium ion	镁离子
precipitate [prɪˈsɪpɪteɪt]	沉淀
dyestuff [ˈdaɪstʌf]	染料
sodium chloride [ˈklɔːraɪd]	氯化钠
accelerant [ækˈselərənt]	促染剂
leveling [ˈlevəlɪŋ] agent	匀染剂
cationic [kætaɪˈɒnɪk]	阳离子的
fixative [ˈfɪksətɪv]	固色剂
reactive [rɪ(ː)ˈæktɪv] dye	活性染料
active group	活性基团
hydroxyl [haɪˈdrɒksɪl] group	羟基

amino [ˈæmɪnəʊ] group	氨基
covalent [kəʊˈveɪlənt] bond	共价键
brightness	明亮度
vat dye	还原染料
reduction	还原
carbonyl [ˈkɑːbənɪl] group	羰基
aqueous [ˈeɪkwɪəs] solution	水溶液
leuco-compound [ˈljuːkəʊˈkɒmpaʊnd]	隐色体化合物
oxidize [ˈɒksɪˌdaɪz]	使氧化
leuco [ˈljuːkəʊ] dye	隐色体染料
sodium salts	钠盐
sulphuric [sʌlˈfjʊərɪk] ester [ˈestə]	硫酸酯
dissolve [dɪˈzɒlv]	溶解
sulphur [ˈsʌlfə] dye	硫化染料
sulphidation [ˌsʌlfɪˈdeɪʃən]	硫化作用
acid dye	酸性染料
aromatic [ˌærəʊˈmætɪk]	芳族的
sulphonic [sʌlˈfɒnɪk] acid	磺酸
disperse [dɪsˈpɜːs] dye	分散染料
non-ionic dye	非离子型染料
cationic dye	阳离子染料
basic dye	碱性染料
wetting agent	润湿剂
softener	柔软剂
dispersing agent	分散剂
dye-carrier	染色载体
sequestrant [sɪˈkwestrənt]	多价螯合剂
dye-fixing agent	固色剂
detergent [dɪˈtɜːdʒənt]	洗涤剂
anti-static agent	抗静电剂
anti-foaming agent	消泡剂
rot-proofing agent	防腐剂

moth-proofing agent	防蛀剂
emulsifier [ɪˈmʌlsɪfaɪə]	乳化剂
mordant [ˈmɔːdənt]	媒染剂
jig dyeing machine	卷染机
guide roller	导布辊
winch [wɪntʃ] dyeing machine	绞盘式绳染机
dye vat	染槽
submerge [səbˈmɜːdʒ]	浸没
winch	绞盘
pad dyer	轧染机
pad-roll system	辊式轧染系统
two or three-roller padder	二辊或三辊式轧液机
infrared [ˈɪnfrəˈred] heating channel	红外线加热通道
reaction chamber	反应室
impregnate [ˈɪmpregneɪt]	浸渍,使充满
open-width washer	平洗机
soaping	皂洗
rinsing [ˈrɪnsɪŋ]	水洗
high temperature and pressure winch beck	高温高压绞盘式绳染机
high temperature and pressure overflow dyeing machine	高温高压溢流染色机
high temperature and pressure jet dyeing machine	高温高压喷射染色机
replenisher [rɪˈplenɪʃə(r)] for dyeing liquor	染液补给箱
rope washer	绳洗机
neutralization [ˌnjuːtrəlaɪˈzeɪʃen]	中和
centrifugal [senˈtrɪfjʊgəl] hydroextractor	离心式脱水机
vacuum hydroextractor [ˈhaɪdrəʊɪksˈtræktə]	真空脱水机
rotary screen dryer	圆网烘燥机
short loop hanging dryer	短环烘燥机
direct printing	直接印花
discharge printing	拔染印花
resist printing	防染印花
printing paste	印花色浆

paste [peɪst]	糊料
alginate [ˈældʒɪneɪt] paste	海藻酸盐糊料
starch paste	淀粉糊料
adhesive [ədˈhiːsɪv]	黏着剂
emulsion [ɪˈmʌlʃən] paste	乳化糊
discharge	拔色
discharge paste	拔染糊料
reducing agent	还原剂
sodium sulphoxylate-formaldehyde [sʌlˈfɒksɪleɪtˌ fɔːˈmældɪˌhaɪd]	甲醛次氯酸氢钠
resist	防染剂
sublistatic printing	转移印花
flock printing	植绒印花
electrostatic flocking	静电植绒印花
roller printing	滚筒印花
screen printing	筛网印花
inkjet printing	喷墨印花
pressure cylinder	承压滚筒
printing roller	印花辊
furnisher [ˈfɜːnɪʃə] roller	给浆辊
colour box	色浆盘
cleaning doctor blade	色浆刮刀
lint doctor blade	刮绒刀
pressure bowl	承压滚筒
screen	筛网
colouring paste	色浆
photogelatin	感光胶
negative image	负片
interstice [ɪnˈtɜːstɪs]	细隙
flat screen printing	平板式筛网印花
rotary screen printing	圆筒式筛网印花
Computer Aided Design (CAD)	计算机辅助设计

inkjet printer	喷墨印花机
digital [ˈdɪdʒɪtl] inkjet printer	数码喷墨印花机
digital textiles	数码纺织品
Continuous Ink Jetting (CIJ)	连续喷墨式
Drop on Demand (DOD)	按需即滴式
piezoelectric [paɪˌiːzəʊɪˈlektrɪk] vibrator [vaɪˈbreɪtə]	压电式谐振器
charge electrode [ɪˈlektrəʊd]	充电电极
deflection [dɪˈflekʃən] electrode	偏转电极
collecting gutter [ˈɡʌtə]	回收槽
electromechnical [ɪˌlektrəʊmɪˈkænɪkəl] transfer method	机电转换法
pulsed signal	脉冲信号
electric thermal transfer method	电热转换法
nozzle clogging [ˈklɒɡɪn]	喷嘴堵塞
pressure regulator [ˈreɡjʊleɪtə]	压力调节器
ink supply pump	供墨泵
recycling [ˈriːˈsaɪklɪŋ] pump	循环泵
cyan [ˈsaɪən]	青色
magenta [məˈdʒentə]	品红
CMYK	青色、品红、黄色和黑色的缩写
printing head	打印头
reactive ink	活性染料色墨
acid ink	酸性染料色墨
disperse ink	分散染料色墨
pigmented [ˈpɪɡməntɪd] ink	颜料型色墨
T-shirt	T 恤衫,短袖圆领针织衫
sweatshirt [ˈswetˈʃɜːt]	长袖圆领运动衫
polo shirts	短袖开领针织衫
apron	围裙
pre-shrinkage treatment	预缩处理

brushing	刷毛
fleece fabric	绒布/纬编针织绒布
napping	拉毛
raising	起绒
cropping [ˈkrɒpɪŋ]	剪绒
calendaring [ˈkælɪndərɪŋ]	轧光
schreinering [ˈʃraɪnərɪŋ]	电光整理
fulling	缩绒
crease resistant finishing	防皱整理
soft finishing	柔软整理
water repellent finishing	拒水整理
peaching	仿桃皮绒整理
anti-static	抗静电
rot-proofing	防腐
moth-proofing	防蛀
sand washing	砂洗
garment making-up	成衣
bath-ratio	浴比
listing	布边色差
tailing	头尾色差
coloured spots	色斑
dyeing unevenness [ˈʌnˈiːvənɪs]	染色不匀
dose	配料
pre-programmed	程序预设的
clour fastness to light	光照色牢度
clour fastness to laundering	机洗色牢度
clour fastness to rubbing	摩擦色牢度
clour fastness to weathering	耐气候色牢度
clour fastness to perspiration [ˌpɜːspəˈreɪʃən]	耐汗渍色牢度
dye uptake [ˈʌpteɪk]	染料上色率
yarn slippage [ˈslɪpɪdʒ]	纱线滑移
pill resistance	抗起球性

angle of crease recovery 折皱回复角

water repellency [rɪˈpelənsɪ] 拒水性

Exercises

1. Dyeing may be carried out ________. ( )
   a) during spinning of cotton yarn
   b) on wool fibres before spinning
   c) after woollen knitwear is made-up
   d) on cotton yarns before weaving
2. Which of the following ways removes wheat starch size better with less damage to fabrics? ( )
   a) Enzyme desizing b) Alkali desizing
   c) Acid desizing
3. The purpose of scouring is to remove ________ on the fibres before dyeing and finishing processes on the fabrics are performed. ( )
   a) waxes b) pectin products
   c) natural colour d) oil emulsion
4. In normal scouring ________ is applied under heat and pressure. ( )
   a) alkaline liquors b) sodium silicate
   c) sodium hypochlorite d) hydrogen peroxide
5. The main reason to avoid using equipment made from iron in bleaching with sodium hypochlorite is because ________. ( )
   a) equipment made from iron rusts
   b) iron is a catalytic agent that will decompose the bleaching agent
   c) equipment made from iron is too heavy to handle
   d) equipment made from iron is difficult to routinely maintain
6. When mercerization is performed, ________ must be applied. ( )
   a) caustic soda b) tension
   c) heat d) hydrochloride acid
7. Which of the following fibres can be dyed with direct dyes? ( )

a) Polyester b) Cotton
c) Wool d) Viscose

8. Which of the following dyes is used in dyeing polyester fabrics? ( )
   a) Reactive dyes b) Sulphur dyes
   c) Acid dyes d) Disperse dyes
9. Which of the following machines is suitable for dyeing polyester knitted fabrics? ( )
   a) Jig dyeing machine
   b) Normal winch dyeing machine
   d) Pad dyer
   d) High temperature-pressure overflow dyeing machine
10. Which of the following printing equipment is simplest to prepare for production? ( )
   a) Roller printer b) Rotary screen printer
   c) Flat screen printer d) Inkjet printer
11. Four colours are usually used in inkjet printing; these are ________. ( )
   a) red, yellow, blue, and white
   b) red, blue, green and yellow
   c) cyan, magenta, yellow and black
   d) red, yellow, blue and black
12. Which of following methods of printing is the most suitable for very large orders? ( )
   a) Roller printer b) Rotary screen printer
   c) Flat screen printer d) Inkjet printer
13. One of the disadvantages for screen printing is that ________. ( )
   a) CAD systems cannot be used in the design preparation
   b) screens need to be changed and cleaned frequently during production
   c) it is not suitable for small orders
   d) it is not suitable for printing stretch fabrics
14. Mercerization can be applied to ________ fabrics. ( )
   a) polyester b) cotton
   c) nylon c) linen

15. Desizing is necessary as a preliminary treatment for ________. ( )
   a) grey cotton woven cloth
   b) grey warp knitted cloth
   c) grey cotton weft knitted cloth
   d) stitch-bonded non-wovens

Reading Materials

Textile Dyeing Effluent and Its Impact on Environment

The textile industry releases significant amounts of colors into water bodies, posing serious environmental issues. It is believed that 12 ~ 15 percent of these dyes are released in effluents during manufacturing processes, causing contamination in the environment. So this industry has a direct link to environmental issues that must be addressed publicly and thoroughly. And it is mainly responsible for an extensive list of environmental impacts.

Textile dyes degrade the aesthetic quality of water bodies by increasing biochemical and chemical oxygen demand (BOD and COD), impairing photosynthesis, inhibiting plant growth, entering the food chain, providing recalcitrance, bioaccumulation and potentially promoting toxicity.

The majority of colors used in the textile industry are light-stable and non-biodegradable. They also have a high resistance to aerobic digestion. Dyes are usually synthetic and have complex aromatic molecular structures, making them more stable and difficult to break down.

— excerpted from *Textile Today*, January 2022, page 48

【参考提示】

1. *Textile Today* 是孟加拉国 Amin & Jahan Corporation Ltd. 出版的关于纺织、服装及时尚类的单月刊杂志。

2. dyeing effluent，染色污水。

3. biochemical oxygen demand (BOD)，生化需氧量。

4. chemical oxygen demand (COD)，化学需氧量。

5. photosynthesis,光合作用;light-stable,光稳定的(光辐射下不易降解)。

6. providing recalcitrance, bioaccumulation,指"造成有害物质降解的抗性和在生物体内的积累"。

7. aerobic digestion,好氧消化(通过好氧菌降解分化)。

ColorJet Launches SubliXpress Plus Industrial Sublimation Printer

ColorJet Launches SubliXpress Plus — 9000 Sq. Meters Daily Production Industrial Sublimation Printer.

Following the success of SubliXpress, ColorJet Group is all set to take transfer paper printing to the next level. Dye sublimation printing is dominant in the digital textile printing sector accounting for almost 50% of all digital textile prints across the polyester & apparel domain. The dye sublimation textile printing market which was valued at USD 8.1 billion in 2019 is facing a CAGR of 10.7% during the forecast period of 2020—2025 and is expected to reach a value of USD 14.2 billion by 2025. Considering the need for a high production sublimation printer to mitigate this high demand, Colorjet has launched SubliXpress Plus.

The SubliXpress Plus comes with 8 staggered Kyocere heads that can print variable drop sizes of 5, 7, and 12 pl as demanded to produce exceptionally sharp lines and create smooth gradients in color tones. The high-speed, high-quality printing with minimal link consumption is a result of decades of experience in manufacturing digital inkjet printers.

— excerpted from *Journal of the Textile Association*, September-October 2021, page 181

【参考提示】

1. *Journal of the Textile Association* 是印度纺织协会出版的双月刊。

2. ColorJet,印度知名的纺织品数码喷墨打印设备制造商。

3. sublimation printing,热升华(转移)印花。

4. CAGR,即 Compound Annual Growth Rate,复合年均增长率。

5. 8 staggered Kyocere heads,8 个交错配置的 Kyocere(国内有译为"京瓷"的)品牌的打印头。

6. pl,皮升,即微微升,10^{-12}升。

Adopting Digital Pigment Printing Technology

Digital printing developments in various dye classes present challenges to adoption by traditional textile printers.

Digital Printhead Solution

The Colaris digital printing technology utilizes the Fujifilm Dimatix StarFire™ industrial printhead. Originally developed for the ceramics industry, the head features 1,024 nozzles per head, the RediJet™ ink circulation system and an integrated heating system. The printhead's modular design makes it repairable with the changing of individual components. In order to add value and extend the system's useful life, Zimmer opened the Colaris Printhead Reconditioning Center offering a repair service for nominal fee.

Interestingly, Zimmer's technology features an "open ink system", which allows the printing company to choose its ink or dye supplier. Zimmer does test inks and dyes for performance in its printheads and system and can make recommendations to Colaris users.

Colaris Pigment Printers

The StarFire industrial printhead currently is the centerpiece of a family of seven Colaris pigment printers. The models range from print widths of 74 millimeters (mm) up to 3,400 mm, and feature square meter per hour (sqm/h) performance at 400 × 800 dpi from 140 up to 1,060 sqm/h. The Colaris family features six-color machines with a maximum 2 to 16 heads per color, depending on the model.

— excerpted from *Textile World*, November/December 2021, page 33

【参考提示】

1. Fujifilm Dimatix,位于美国的富士集团旗下的知名喷墨打印头供货商,StarFire™是它的打印头注册商标。RediJet™为打印头采用的墨水循环技术的注册商标名称。

2. Zimmer,奥地利知名筛网印花及数码印花设备制造商。

Keys to Exercises

Chapter 1

1. b/d 2. b 3. c 4. b/c/d 5. b/d 6. b 7. a/c 8. a/b 9. a/b/c
10. a/c

Chapter 2

1. b 2. d 3. c 4. d 5. b/c 6. c 7. b/c/d 8. a/c 9. b 10. d

Chapter 3

1. a/c/d 2. d 3. c 4. b/d 5. d 6. b/c 7. c 8. d 9. b 10. d

Chapter 4

1. b/d 2. d 3. c 4. b/d 5. b 6. c 7. b 8. d 9. b 10. a/b/c

Chapter 5

1. a/d 2. c 3. d 4. c/d 5. a/c/d

Chapter 6

1. b/c/d 2. a 3. a/b/d 4. a 5. b 6. a/b 7. b/c/d 8. d
9. d 10. d 11. c 12. a 13. b 14. b/c 15. a

Chinese Version
中文参考译文

第 1 章　纺织纤维

纤维是纺织品的基本元素。一般来说,直径从几微米到几十微米,长度比细度大许多倍的材料可以认作为纤维。纤维之中,长度数十毫米以上并且具有足够的强度和柔韧性的纤维可以归类为纺织纤维,用以生产纱、绳或织物。

1　纺织纤维的种类

纺织纤维的种类很多,但所有纺织纤维都可以被分为天然纤维和化学纤维。

1.1　天然纤维

天然纤维包括植物纤维、动物纤维和矿物纤维。

从受欢迎的角度而言,棉花是最常用的植物纤维,亚麻和苎麻紧随其后。亚麻很常用,但因为亚麻纤维较短(25 ~ 40 毫米),传统上将棉花或涤纶纤维和它混纺。苎麻即所谓的“中国草”,是一种耐用的且带有丝质光泽的韧皮纤维,它具有极好的吸着性,但制成的织物容易起皱褶,因此苎麻常常和合成纤维混纺。

动物纤维或者来自动物的毛发,比如,羊毛、羊绒、马海毛、驼毛和兔毛等;或者来自动物的腺分泌物,比如,桑蚕丝和柞蚕丝。

最常听说的天然矿物纤维是石棉。石棉是一种防火性极好的无机纤维,但它也有害于健康,因此现已不使用。

1.2　化学纤维

化学纤维可以分为有机纤维或无机纤维。前者又可以分为两类,一类通过天然聚合物的转化,形成有时人们所称呼的再生纤维;另一类由合成聚合物制取,形成合成长丝或纤维。

常用的再生纤维有铜氨纤维(CUP,通过铜氨法获取的纤维素纤维)和黏胶纤维(CV,由黏胶法获取的纤维素纤维。铜氨纤维和黏胶纤维都可以称为 rayon)。醋酯纤维(CA,醋酯纤维素纤维,其中 74% 至 92% 的羟基被乙酰化)和三醋酯纤维(CTA,醋酯纤维素纤维,其中至少 92% 的羟基被乙酰化)是另一类再生纤维。Lyocell (CLY)、Modal (CMD)以及 Tencel 是当今流行的再生纤维素纤维,它们的研发是为了满足生产中对环境问题的要求。

如今,再生蛋白质纤维也日益受到欢迎,它们之中有大豆纤维、牛奶纤维和甲壳素纤维。再生蛋白质纤维特别适合医学上的应用。

纺织上使用的合成纤维通常由煤、石油或天然气制得。通过不同的化学反应,它们的单体经聚合,形成化学结构相对简单的、能够熔融、或能够在合适的溶剂中溶解的高分子聚合物。常用的合成纤维有聚酯纤维(PES)、聚酰胺纤维(PA)即尼龙、聚乙烯纤维(PE)、聚丙烯腈纤维(PAN)、变性聚丙烯腈纤维(MAC)、聚丙烯纤维(PP)和聚氨基甲酸酯纤维(PU)。诸如聚对苯二甲酸丙二酯纤维(PTT)、聚对苯二甲酸乙二酯纤维(PET)和聚对苯二甲酸丁二酯纤维(PBT)之类的芳族聚酯纤维也日益受到欢迎。除此之外,许多具备特别性能的合成纤维业已研制成,其中很知名的有 Nomex、Kevlar 和 Spectra。Nomex 和 Kevlar 是杜邦公司的注册品牌。Nomex 是具有上乘阻燃性能的间位芳族聚酰胺纤维,Kevlar 由于其超常的强度可以用来制作防弹背心。Spectra 由聚乙烯制得,具有超高的相对分子质量并被视为世界上最强且最轻的纤维之一。它特别适用于盔甲、太空产品以及高性能的运动物品。研究仍在不断深入,纳米纤维的研究就是该领域中最热门的话题之一。为了确保纳米粒子对人类和环境的安全性,派生出一个被称为"纳米毒理学"的科学新领域。该领域目前在关注开发用以调查和评估纳米粒子、人类和环境之间交互作用的测试方法。

常用的无机化学纤维有碳纤维、陶瓷纤维、玻璃纤维和金属纤维。它们大多用于特殊目的,以实施特定功能。

2 纺织纤维的性质

对于纺织纤维的性质已有大量的研究,这些研究包括了纤维的吸着性的研究,以观察某特定的纤维是亲水的、疏水的、吸湿的、亲油的还是疏油的;还包括纤维的其他性质方面的研究,如强度、弹性回复、抗磨性、挠性、蠕变、可燃性、化学性质以及抗生物能力等。图 1.1 简要地总结了一些常用纺织纤维的性能特征。

一般来说,蛋白质纤维有较高的回弹性,且为亲水性纤维。它们的机械性能随吸湿而改变。碱能够削弱它们的力学性能,紫外线可以使它们发黄并且强度降低。根据它们特定的形态和化学结构,不同蛋白质纤维的实际性质会不同。比如,羊毛表面具有鳞片层,这使得羊毛易于毡化,

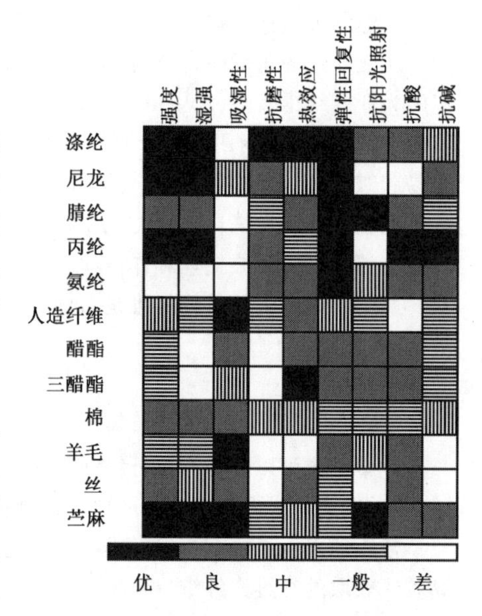

图 1.1　重要的纺织纤维的性能等级

除非它已经防毡化处理。相反，真丝表面光滑，这赋予真丝闪亮的光泽。

纤维素纤维也是亲水性的，它们的机械性能也会随吸湿而变化。与天然蛋白质纤维相比，纤维素纤维的回弹性较逊色但它们的抗碱降解能力强得多。其中，苎麻具备很好的强度及抗紫外线的能力。它们的具体形态和化学结构也会影响它们的性质。单根棉纤维是卷曲的，像瘪了气的水管——棉具有较低的绝热能力，因为大多数棉纤维的纤维腔成熟晒干后干瘪。黏胶纤维的化学成分和棉纤维相似，这使它具备和棉相似的特点。然而，由于黏胶的聚合度和结晶度都较棉低，黏胶的吸湿性较好但拉伸强度较低，特别是在湿态时。

合成纤维一般具有较低的回潮性，它们大多是亲油但疏水的。和蛋白质以及纤维素纤维不同，合成纤维具有良好的抗蛀、抗霉、抗真菌的性质。合成纤维的实际性质将取决于它们的分子长度、化学组成、聚合体的排列、分子间的键以及截面形状等。比如，无定形区越多，分子结构中氢键或极性基团越多，纤维的亲水性就越好；沿纤维轴向排列的分子越多，纤维的强度就越高。为了增加取向度，大多合成纤维在制造中经牵伸。合成纤维的性质还很大程度上取决于它们的化学构成。聚酯纤维的强度高是因为它的结晶度较高，吸湿性差是因为它缺少亲水基团，良好的抗酸性和稍逊色的抗碱性是因为它的化学构成。

3 纤维的质量

必须关注纤维的质量，因为它将严重影响由它制成的纱线和织物的质量。纤维的质量可以从两方面考虑，即，外观质量和内在质量。

生产中纤维的黏结将影响化学纤维的外观质量，此外，纤维表面疵点也可能影响它的内在质量。因未充分去硫而在黏胶纤维上形成的硫斑就是另外的一个例子。

内在质量主要有机械质量和化学质量，它们将可能影响纤维后加工或最终用途。需要测量断裂强度、断裂伸长率、纤维长度偏差、纤维细度均匀度、纤维倍长率、卷曲频率和回潮率等，以评估纤维质量。对某些纤维可能需要进一步测试，比如，黏胶可能需要测试它的湿强、钩接强度和残硫量，腈纶可能需要测试它的上色率，涤纶可能需要测试它的沸水收缩率。

对于天然纤维，进行拉伸测试以确定它们的断裂时的强度和延伸性，以此可以计算它们的断裂强度和伸长的偏差。通常还测量样本纤维的细度以确定它们的平均值和变异系数。诸如棉纤维中的杂草屑和羊毛中的油脂等任何杂质，需要进行检测，它们是纤维品质的评估要素。

测试前，纺织纤维应该进行温湿度调节，使得需测试的材料在标准温湿度测试条件下达到吸湿平衡状态。纺织纤维测试在标准温湿度条件下进行，该条件为在当地的大气压下，相对湿度为65%±2%，温度为20℃±2℃。对于某些材料，诸如已知的相对来说不受相对湿度变化影响的涤纶和腈纶，相对湿度的容差可以扩展到±5%。对于纱线和织物的测试，特别是对于那些对湿度敏感的，同样的条件也应该严格遵守。

第2章 纱

纱是构成针织物或机织物的基本元素。纱可以由短纤维或长丝制得。纤维或长丝的选择以及纱的结构会在很大程度上影响以后由该纱制得的织物的性质。

1 长丝纱

长丝实际上就是非常长的纤维,因而通常称之为长丝。除真丝是天然长丝外,纺织业使用的大多数长丝为化学长丝。仅由单根丝构成的长丝纱被称为单丝,由两根或两根以上的丝构成的长丝纱被称为复丝。针织或机织中常使用复丝,因为它们较软、较韧。

化学长丝或纤维的纺制方法有两种——熔体纺丝和溶液纺丝。诸如涤纶、锦纶或丙纶之类在分解前熔融的聚合物通过熔体纺丝纺制。将聚合物切片熔化,熔融的聚合物被挤压通过喷丝孔,在周围的空气(或水)中固化。大多数这类聚合物切片由石油制得,因此,国际市场石油价格的任何波动会导致相应的化学纤维价格的波动。

溶液纺丝用于那些没有明确熔点的或者在到达熔点前就分解或炭化的聚合物。溶液纺丝可再分为干法纺丝和湿法纺丝。在溶液纺丝时,应先制备纺丝溶液,然后,如为湿法纺丝,从喷丝头挤出的聚合物溶液经过某种液体,丝在其中固化;如为干法纺丝,挤出的聚合物溶液通过热空气(溶剂蒸发)而固化。由于在干法纺丝中,不得不采用更多专门的措施以控制环境污染,它的成本一般高于湿法纺丝。醋酯和三醋酯长丝或纤维由干法纺丝制得,而腈纶和黏胶通过湿法纺丝制得。

通过喷丝头形成的最初的长丝一般不能用于纺织。它们总是要经牵伸,使分子链定向,增加长丝的强度和延伸性。除此,还要进行集束、水洗、上油、卷曲和热定形工序,以使纤维更强并适合纺织上的最终用途。

化学纤维可以有光泽,但如果加入消光剂,比如二氧化钛,可以制成全消光或半消光纤维,这取决于所加入的消光剂的量。

随着制造工艺的发展,随着喷丝头的设计或喷丝头上的喷丝孔形状的改进,各种各样的双组分纤维、超细纤维以及异形纤维现已能够常规生产。

当然,生产长丝时,应关注其质量。长丝在打包发运到针织厂或机织厂前,牵伸、加捻、水洗、热定形以及络筒都可能有要求,最后的那道工序是为了使长丝卷绕成客户所要求的适合以后加工的卷装。

长丝可以经变形加工使其缠绕,并使其更蓬松,延伸性更好。这可以通过对长丝假捻,热定形后再退捻而获取。另一常用的方法是空气变形技术,在该技术中长丝被超喂送入空气涡流中。

上述方法形成的长丝可以切断或拉断成称之为“短纤”的纤维。“短纤”指的是它们的长度较短。当化纤长丝被切断或拉断成短纤维时,为了获得所需纤维长度范围,应该控制好纤维长度。一般,长度在40毫米以下的纤维被称为短纤维或棉型纤维,长度在60至70毫米间的被称为长纤维或毛型纤维,那些长度在40至60毫米间的为中长纤维。

2　短纤纱

短纤纱由短纤制成。短纤包含了诸如棉、羊毛或亚麻等天然纤维,或诸如涤纶、锦纶或腈纶等化学纤维,或它们的混合,这被称为“混纺”纱。

2.1　环锭纺纱

为了将纤维结合,形成具有足够抱合力以确保后道加工所需的足够的强度和延伸性的纱线,纤维需要尽可能地随机配置并且进行加捻。通过加捻成纱的方式称为纺纱。最常用的纺纱工艺可能是环锭纺,它的详细工序某种程度上取决于所涉纤维的种类以及对纱线的最终规格要求。典型的棉纱环锭纺工序如下:

前处理→梳棉(粗梳。编者注)→(精梳)→牵伸→粗纱→细纱

首先,为了生产质量均匀并达具体价格要求的纱线,以打包形式运抵纺纱厂的短纤进行开棉、清棉和混棉。若生产混纺纱,不同种类的纤维可以按预定的比率,通常为重量比,在此阶段混合。接着,纤维由梳棉机的旋转着的滚筒上覆盖的钢丝或锯齿针布分离,部分定向,这将纤维形成粗梳生条(图2.1)。

图2.1为开棉、清棉和混棉后的棉卷如何转化为粗梳生条的一个示例。卷绕在棉卷扦上的棉卷安放在棉卷架上,棉卷罗拉的旋转使棉卷退绕。当棉卷到达给棉板时,给棉罗拉将棉卷推向表面覆有金属锯齿针布的刺辊(英文中刺辊除了称为licker-in,还被称为taker-in)。刺辊高速转动,从棉卷中拉出小股的纤维簇。离心力以及除尘刀的作用使得棉卷中的大多数杂质和非常短的纤维被分离,并且从刺辊下的下壳罩(中文俗称“小漏底”。编者注)的栅格中落下。锡林也覆有金属锯齿针布,并且以非常高的速度转动。锡林上的针布的齿尖和锡林上方循环运动的盖板上的针布的齿尖共同作用开松纤维簇,这个共同作用被称为粗梳。同时,剩下的杂质和短纤维多数被去除,它们中的一些,通过锡林下壳罩(中文俗称“大漏底”。编者注)的栅格落下;一些形成盖板花,这些盖板花由上斩刀和盖板刷帚清除。粗梳后单纤维从锡林转移到转速较慢的也覆有金属锯齿针布的道夫上。剥棉辊将经过粗梳的纤维网从道夫上剥离。在纤维网进入将其聚成粗梳生条的喇叭口前,两个轧辊轧挤出剩余的杂质。最后,生条盘入生条筒。

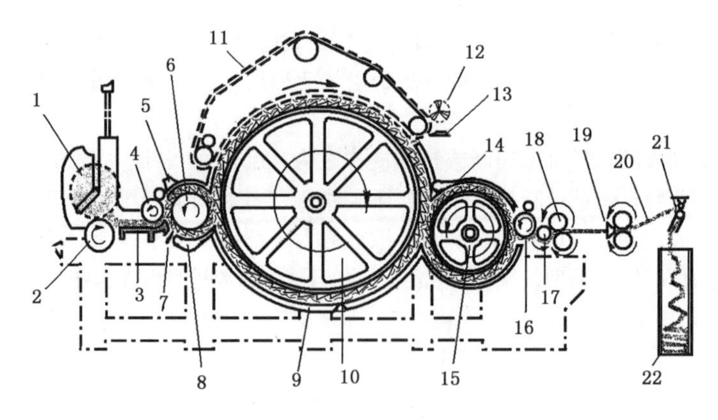

图 2.1　梳棉机

1—棉卷　2—棉卷罗拉　3—给棉板　4—给棉罗拉　5—吸尘罩　6—刺辊　7—除尘刀　8—小漏底
9—大漏底　10—锡林　11—盖板　12—盖板刷帚　13—上斩刀　14—吸尘罩　15—道夫　16—剥棉罗拉
17—转移罗拉　18—轧辊　19—喇叭口　20—粗梳生条　21—圈条器　22—生条筒

前述的粗梳机显示了棉型的或较短的短纤粗梳机的特点。新型的粗梳机上,纤维通过管道喂入刺辊而不是以纤维网喂入。除了循环运动的盖板被换成数组被称为梳毛辊和剥离辊的罗拉外,粗纺毛纱和精纺毛纱的粗梳以类似的原理进行。这些罗拉通常覆有金属锯齿针布,但在较老式的粗梳机上,覆盖的是弹性的钢丝针布。这些罗拉相互作用并和锡林相互作用,开松纤维。粗纺毛纱的粗梳机和生产非织造织物纤维网的粗梳机有两个锡林和与它们配合的梳毛辊和剥离辊。

在牵伸工序,数根粗梳生条在并条机(图 2.2)上被牵拉成一根纱条,以获得更好的混棉效果。对于混纺纱,不同种类纤维的生条也可以在此阶段混合,以满足更精确的混纺比的要

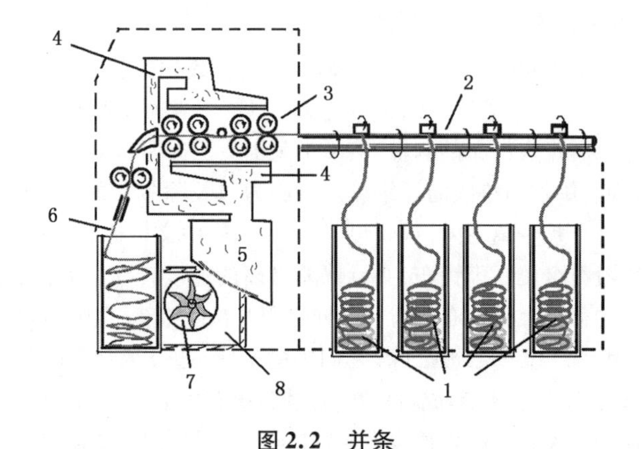

图 2.2　并条

1—生条　2—导条罗拉　3—牵伸装置　4—吸尘器　5—滤尘箱　6—熟条　7—排风扇　8—排风管

求。牵伸至少做两次，这有助于纤维排列更平行，减少前弯钩纤维和后弯钩纤维的数量。因为纤维要通过并条机，每次牵伸被称为“一道”。因此，经两次牵伸的纱条将谓之二道熟条。然后，熟条被送到粗纱机（图2.3）以生产粗纱。粗纱就是短纤的极长的集合体，短纤充分地平行排列，稍有加捻，在细纱纺纱准备的后阶段还能够被牵伸。

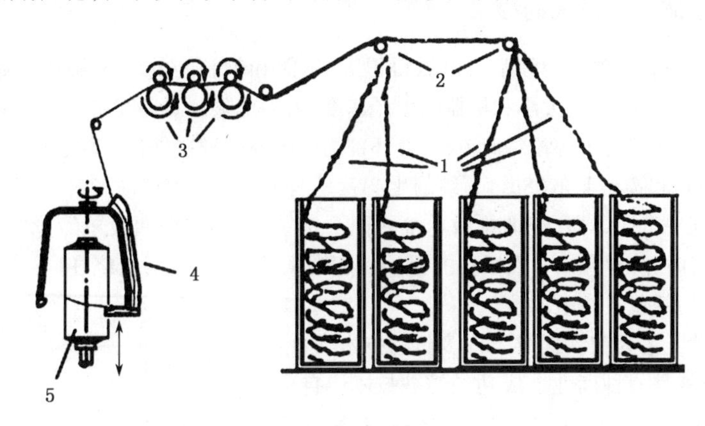

图2.3 粗纱

1—熟条 2—导条辊 3—牵伸装置 4—锭翼 5—粗纱筒子

在纺纱的最后工序中，粗纱拉伸变细以达所需的线密度（单位长度的重量），这是细纱纺纱准备的最后牵伸。用环锭纺方法加入所需捻度。在环锭纺工序，装在锭轴上的纱筒转动，对来自机器牵伸系统前罗拉的纤维加捻。纱穿过随锭子转动而绕钢领滑动的小钢丝圈，由于施加在纱上的拉力，加捻后的纱卷绕在卷装上（见图2.4）。

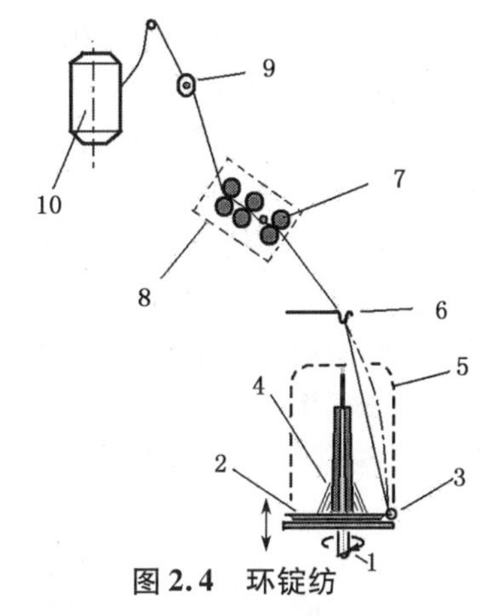

图2.4 环锭纺

1—锭轴 2—钢领 3—钢丝圈 4—纱筒 5—隔纱板 6—导纱钩 7—前罗拉 8—牵伸装置 9—张力器 10—粗纱

如果需要更好质量的纱，牵伸前粗梳生条可以经过精梳机，进一步将纤维拉直，并且去除短的纤维和杂质，这样能够生产较细的纱。通过附加的精梳工序生产出来的纱被称为精梳纱，未经精梳的纱被谓之普梳纱。精梳纱用以生产较薄的质量较好的织物。

在粗纺毛纱时，直接在梳毛机生成的是重量轻且连续的短纤纱条状的头道粗纱，而不是粗梳生条，然后，头道粗纱通常使用环锭纺机或走锭纺机直接纺纱。

精纺毛纱工序和精梳棉纱工序相似。它经粗梳、针梳（和牵伸相似的一个工序，它在纤维牵伸时使用针梳梳理纤维）、精梳、再针梳，生成粗纱，然后细纺。粗纺毛纱和精纺毛纱工序并不仅限于羊毛纤维，它们可以用于那些和正常情况下采用这些方法纺纱的毛纤维的长度和细度都相仿的合成纤维。通常，经过精梳的且使用精纺毛纱工序生产的长纤维的短纤纱被称为精纺毛纱，而未经精梳的且使用粗纺毛纱工序生产的长纤维的短纤纱被称为粗纺毛纱。精纺毛纱工序如下：

前处理→粗梳→前道针梳→精梳→后道针梳→细纱→后处理

2.2 其他纺纱工艺形成的纱

环锭纺有一些局限。钢领不可以做得很大，钢圈和钢领间的摩擦限制了锭轴的转速。这制约了可以纺纱的卷装大小和机器的生产速度，大大地限制了环锭纺方法的发展。

然而，由于环锭纺机构相对简单，开发环锭纺的研究已经在进行。基于环锭纺技术的赛络纺（Sirospun）和赛络菲尔（Sirofil）纱的生产就是两个例子。生产前者时，来自前罗拉的两股粗纱穿过同一个导纱器，经加捻后卷绕在同一锭子上，形成和两股单纱捻向相同的双股纱。生产后者时，长丝和来自前罗拉的粗纱一起喂入，两者加捻后卷绕在同一锭子上，形成以长丝为芯、短纤包覆其上的纱。

现代纺纱工艺为纺纱提供了许多新的技术，自由端纺纱（OE）就是一种商业上成功的纺纱技术。自由端纺纱系统中取消了环锭纺纱生产中使用的粗纱工序。通常，二道熟条喂入自由端纺纱机中，纤维散开后和正在生产的纱的自由端结合。赋予纱强度的加捻在此时实施。常用的自由端纺纱类型有转杯纺纱和摩擦纺纱。在转杯纺纱中，通过将纤维配置在转杯的内壁并通过位于转杯中央的气嘴将纱拉离转杯壁，纱得以形成并被加捻（见图2.5）。槽筒的主动回转使纱筒转动，在沟槽的导引下，纱沿着纱筒的轴向来回运动，卷绕在纱筒上。

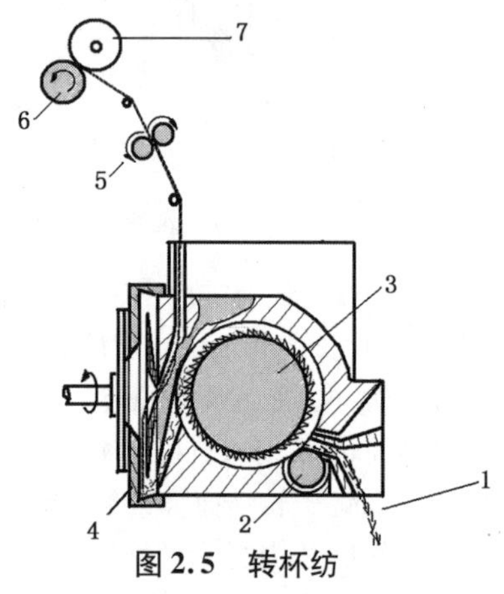

图2.5 转杯纺

1—熟条 2—喂给罗拉 3—分梳辊 4—转杯 5—输出罗拉 6—槽筒 7—纱筒

在摩擦纺纱中，纤维被置于一对间隔很近的摩擦滚筒之间，至少有一个滚筒为排有吸孔的抽吸滚筒。通过将纤维沿滚筒轴向拉离滚筒进行纺纱，随滚筒转动，通过作用在纤维上的摩擦力，滚筒对纱加捻（见图2.6）。除转杯纺纱和摩擦纺纱外，市场上还可以看到喷气纺纱和涡流纺纱生产的纱。由于对纱的加捻和卷绕是分开的，自由端纺纱可以纺制较大的纱筒。另外，转动纱的小小的自由端要比像环锭纺那样转动整个纱筒方便得多，这使得自由端纺纱产量较高，并且能源也节省。

另一种受欢迎的用于长纤维精纺毛型纱的纺纱方法为自捻纺纱。自捻纺时，两根粗纱被牵伸后通过一对旋转且做轴向往复运动的罗拉（见图2.7）。往复运动对须条交替施加S向捻和Z向捻。两股须条合在一起并相互捻合，生成沿长度上交替S捻和Z捻的双股自捻纱（ST）。这类纱以后可以再经加捻，形成加捻自捻纱（STT）。这种纺纱方法的生产率远高于环锭纺，然而，纱中加捻的变化会在某些织物上形成不想要的图案效应。

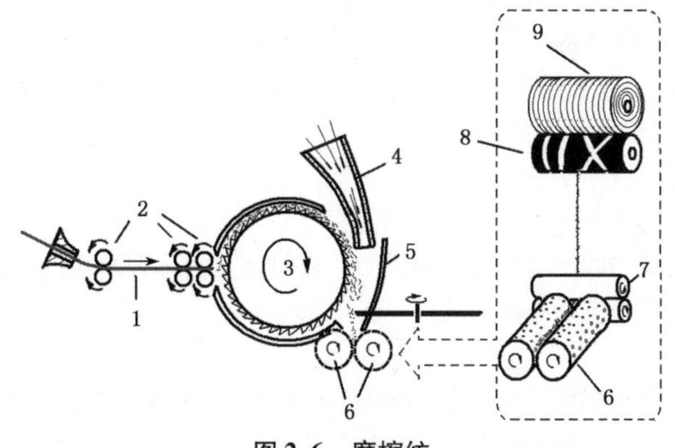

图2.6　摩擦纺

1—熟条　2—牵伸装置　3—分梳辊　4—吹风管
5—挡板　6—抽吸式滚筒　7—输出罗拉　8—槽筒　9—纱筒

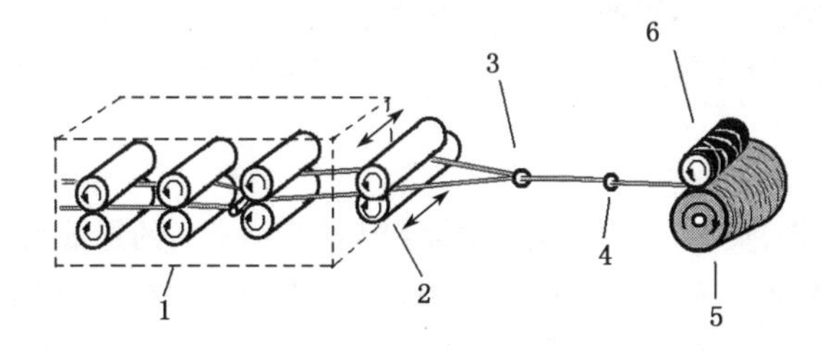

图2.7　自捻纺

1—牵伸装置　2—搓捻辊　3—汇合导纱钩　4—导纱钩　5—纱筒　6—槽筒

2.3　股纱

将两股或多股单纱捻在一起可以得到股纱或线。研究表明,加捻过程中所涉及的单纱根数应该小于五,否则股纱的结构将不稳定。双股纱常用于机织或针织,但缝纫线一般采用三股结构。生产股纱时,股纱的加捻方向一般和构成它的单纱的捻向不同。通常,S 捻的单纱 Z 向加捻,形成股纱。

3　纱的质量

可用多种方法对纱分类,比如,按是否在纺纱时混有两种或两种以上的纤维分为纯纺纱或混纺纱;按是否涉及精梳工序分为普梳纱和精梳纱;按纱的形态或结构分为单纱、股纱、单丝、复丝、膨体纱、变形纱、包覆纱、包芯纱和竹节纱等;按采用的纺纱技术分为环锭纺纱、赛络纺纱和自由端纺纱(大多数为转杯纺纱)等。

随着纺纱工艺的发展和纺织原材料的研制,市场上出现越来越多的种类的纱。

纱的质量是一个重要的问题。纱的质量可从许多方面来评估,这可包括纱的线密度、线密度偏差或不匀率、捻度和捻度不匀率、纱的强度,特别是拉伸强度和钩接强度、断裂伸长等。此外,还需要观察和评估表面纱疵。

3.1 纱的线密度

纱的线密度是关于纱的细度规格的说明。它的计算方法有两种:定长制和定重制。用相同的方法可以确定纺织纤维的细度。

3.1.1 定长制

特克斯是定长制下国际标准化组织(ISO)定义的纱线线密度的标准单位。特克斯等于 1 000 米长的纱在公定回潮率下的克重。对于较细的纱、长丝甚至纤维,常使用分特克斯(dtex),它等于特克斯的十分之一。

旦尼尔为特克斯的九分之一,传统中也用于定义长丝的线密度。即,一根长丝的旦尼尔等于 9 000 米长的该长丝在公定回潮率下的克重。

可见,在定长制下特克斯或旦尼尔的值越大,线密度将越高,对于给定的纤维密度,纱就越粗。

3.1.2 定重制

英制棉纱支数(即,英支)是国际贸易中棉纱或棉型纱常用的定重制下的单位。英支数被定义为公定回潮率下一磅重的纱所含的 840 码长的倍数。粗纺毛纱支数和精纺毛纱支数也是在类似的基础下定义的,只是所涉码数不一样,前者采用 256 码,而后者采用 560 码。

和英支相似的单位为公支,它被定义为公定回潮率下一公斤的纱所含的 1 000 米长的倍数。

在定重制下,支数越小,纱的线密度越高,纱就越粗。

如果两股单纱,比方说 14 tex,加捻成股纱,该股纱的线密度表达成 14 × 2 tex;但由两根单纱,比方说每根 42 英支,形成的股纱的线密度被表达成 42/2 英支。对于复丝,需确定它的线密度和丝的根数。比如,75 dtex/30F 表示一根复丝,其总线密度为 75 分特,含有 30 根单丝。

3.2 纱的捻度

为使纱内的纤维之间产生抱合压力,需要加捻。这样,当施加沿纱的轴向的拉力时,纤维表面产生摩擦力,而摩擦力又减少了纤维的滑移,同时赋予纱一定的强度和延伸性。加捻可以有左手捻(S 向捻)或右手捻(Z 向捻)。

捻度表示加捻的程度,它定义为每单位长度的纱中的捻数或捻回数。单位长度的捻回数越多,纤维间的抱合力越大,直至一临界点。捻度再增加并超过给定值,将会增加纤维在纱内的螺旋角(即捻角),这样,每根纤维的断裂强度就不能有效释放,使得纱实际的强度减弱。该给定值取决于纤维的类型、密度和纱的结构与密度。

捻度按施加在一定长度(比如一米)纱上的捻数而测得,从中可以定义捻系数,它由上述捻度和纱线线密度导出。捻度和捻向很重要,在订购纱的时候需要说明,比如将一米长度内有 S 捻向、640 捻数并且线密度为 40 特克斯的单根纱的规格定义为 40tex S640。

3.3 纱的强度和伸长

纱的强度,尤其拉伸强度和相应的断裂伸长,是需要控制的重要力学参数。纱的断裂由纤维滑移和断裂造成,因此纤维的特性、纱的线密度和捻度影响纱的强度。其次,由于纱在最弱处断裂,线密度或捻度不匀率越大,纱就越弱。此外,纱线密度偏差可能会在织物上显示,比如形成条花,尤其这种偏差是周期性的,由此影响织物的外观及其机械性质。

纱为具体的用途而纺制,因此相应的质量要求会不同。例如,针织用纱应该较机织用纱柔韧,因此趋于采用较低的捻度。此外,圆针织机用纱的捻向也需要小心加以选择。

对于功能性纱线,必须进一步观察某些特别的性能。例如,对于用于生产智能纺织物的导电纱线,可能不得不测取和评估诸如电阻之类的电气性能。

买方在购买纱线时应该明确规定其精确的要求,规定纱的类型和带有容差的详细规格。纱的规格应该包括纤维的类型、混纺率(如要求混纺纱的话)、线密度、捻度(或捻系数)以及捻向。此外,所采用的纺纱技术、断裂强度、不匀率的限制等等也可能需要说明。

供货方应该评估买方的要求是否可行,如有必要,应该向买方弄清纱的规格容差,因为生产具有确定且绝对量值的捻度或断裂强度等方面的纱是不可能的。

第3章　机织和机织物

1　机织织机基本特征

人们相信，自公元前4 400年就有织机在使用。多少年来，机织工艺和机织织机经历了极大的发展。然而，无论是现代的或者是传统的织机，它们都具有相同的基本特征。织机上都会有送经机构、开口机构、引纬机构、打纬机构和织物牵拉卷取机构（图3.1）。

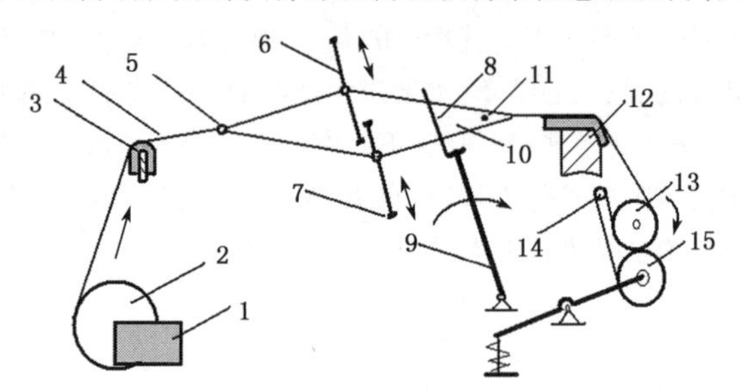

图3.1　机织基本图示

1—送经机构　2—经轴　3—后梁　4—经纱　5—分经棒
6—综丝　7—综框　8—钢筘　9—筘座　10—开口
11—纬纱　12—胸梁　13—牵拉辊　14—导布辊　15—卷布辊

送经机构逐步松开卷在经轴上的经纱，使之向织造区域运动。传统织机使用消极送经系统，经纱由主要因织物牵拉机构产生的来自织造区域方向的力拉离经轴。现代织机使用积极送经机构，经纱的送经速度通过机械的或电子的手段被积极控制和改变。通过安装在胸梁处的传感器测出经纱张力，来自传感器的信号通过控制电路调整送经和牵拉电机。

开口机构是用以分离或打开各组经纱（或通常所谓的经纱片）形成通道（即开口）以便纬纱通过的机构。在织机上，综框内的综丝均匀配置。每根综丝有一综眼，各根经纱从中穿过。当一综框向上运动时，穿过该综框内综丝的经纱将上升；反之，当综框向下运动时，该经纱将下降。最简单的织机使用2片综框，它们交替上下运动，形成开口。可见，一台织机的综框数越多，该织机能够生产的结构就越复杂。一台典型的简单织机装有由凸轮或踏综盘控制的2至8片综框。如果采用“多臂”装置控制综框的运动，织机可能有16至32片综框。

这些综框通过打孔纸板链编程,或如今更常见的,通过电子控制系统编程,以生产更复杂的结构。如果使用贾卡提花机构,每根经纱的升降能够单独控制,不使用综框。这样能够生产非常复杂且花型循环非常大的花型。

引纬机构将纬纱通过开口引入。较老式的传统织机使用梭子。当开口形成,打梭棒击打梭箱内的梭子。在梭织前,纬纱先绕在纡子上。为了将纬纱引入织口,梭子携带绕有纬纱的纡子,穿过开口,并由织机另一侧的梭箱接纳。质量较大的木制梭子在开口间的缓慢往复运动以及梭子有限的携纱能力大大阻碍了生产率,并且,纬纱的张力也难以控制。此外,打梭和接梭产生织造噪音。因此,大多数现代工厂采用的是诸如剑杆织机、喷气或喷水织机、片梭织机之类的无梭织机。和梭织机不同,这些织机很大的优点在于引纬时不必在开口间运送纱纡。

剑杆织机使用单剑杆或双剑杆(送纬剑杆和接纬剑杆,见图3.2),这取决于织机的宽度。剑杆有挠性的或刚性的。在较常见的双剑杆织机上,当开口形成,两根剑杆都进入开口。送纬剑杆握持纬纱,并且将纬纱从织机一侧静止的纱筒上拉出。当它们在织物横向中部相遇,接纬剑杆从送纬剑杆处接过纬纱,然后,两剑杆回退。接纬剑杆的回退完成引纬。

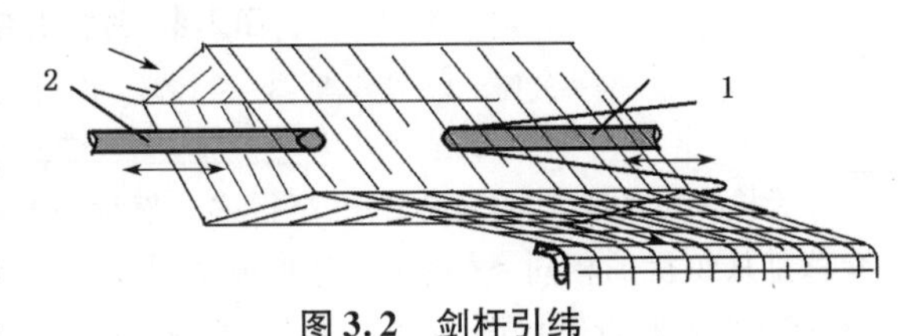

图3.2 剑杆引纬

1—送纬剑杆 2—接纬剑杆

在喷气织机上,纬纱由喷射的气流携带穿过开口(图3.3)。来自静止的纱筒上的纬纱经过纱线张力器,进入测纬和储纬装置。该装置测取所需长度的纬纱并使之保持在松弛状态。然后,纬纱被引入喷嘴。位于测纬和储纬装置与喷嘴间的纱夹握持纬纱。当开口形成,纱夹松开纬纱,喷嘴的高压气流将纬纱从储纬装置中拉出,使之穿过开口。纬纱引入后,纱夹再度握持纬纱。位于喷嘴前的夹剪剪断纬纱,为下一个引纬做好准备。在某些机器上,为了帮助纬纱横穿开口,沿织机横向装有辅助喷嘴。某些喷射织机利用喷射的水来迫使纬纱通过开口。

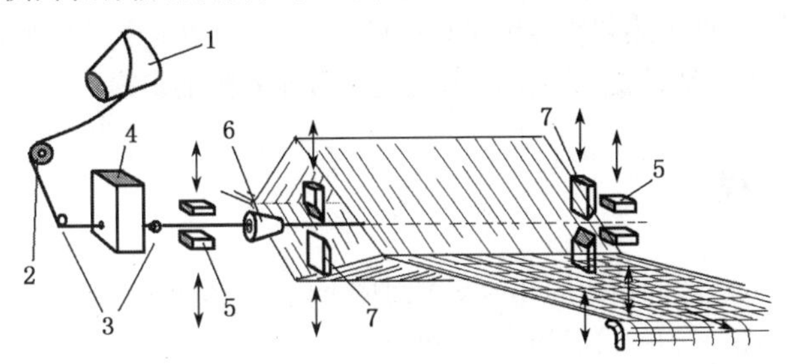

图3.3 喷气引纬

1—纱筒 2—张力器 3—导纱器 4—测纬储纬装置 5—纱夹 6—喷嘴 7—纱剪

片梭织机使用片梭(图3.4)。和前述两类织机一样,片梭携带单根纬纱而不是纡子穿过开口。携带纬纱的片梭射出,穿过开口,然后在布的下方往回输送,准备以后的引纬。

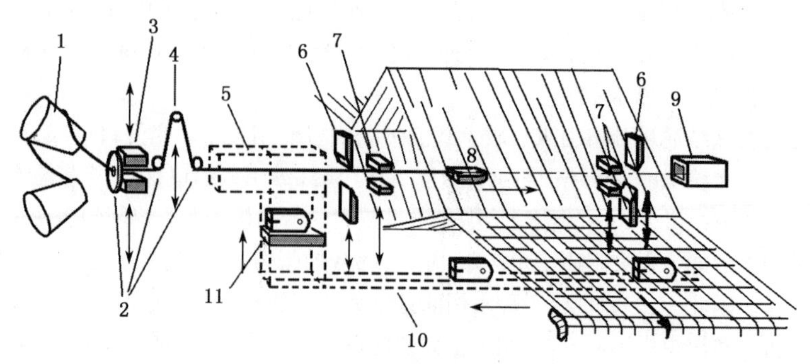

图 3.4　片梭引纬

1—纱筒　2—导纱器　3—夹纬器　4—张力杆　5—输纬器和输梭器
6—纱尾剪　7—纱尾夹　8—片梭　9—制梭器　10—传梭装置　11—提梭器

无梭织机的生产率较梭织机高得多。然而,由于纬纱总是从织机的相同一侧引入,这形成了需要裁去的毛边,或需要安装折边装置,在引纬后将纱尾折入到织物中(图 3.5)。

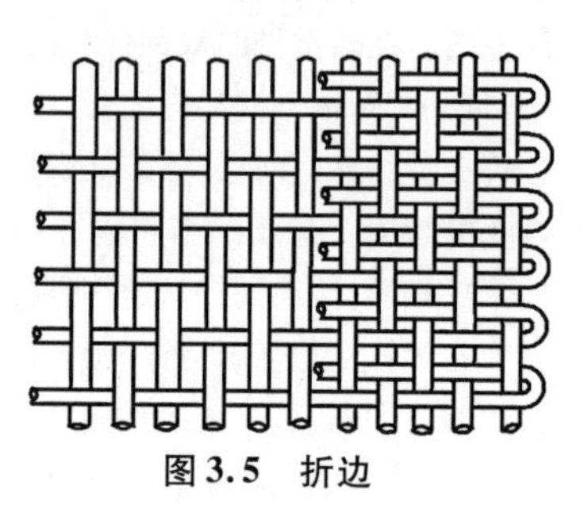

图 3.5　折边

打纬装置将引入的纬纱打入织口。该装置包含安装在筘座上的梳状钢筘。所有经纱,通常成组,穿过筘片间谓之筘隙的均匀间隔的间隙。当每根纬纱引入,筘座向前摆动,钢筘将纬纱紧紧地推向织口,形成织物。

牵拉卷取装置包含一系列的辊子。牵拉辊的积极旋转将织物拉离编织区域,织物被卷上卷布辊。牵拉力还是使经纱绷紧的主要因素。经纱和织物的张力、送经和牵拉的速度以及纬纱的线密度决定了织物的纬密。

在现代织机上,某些上述装置是电子控制的。比如,采用电子开口装置,各片综框的停留和交错时间可以方便地设置,从而高效率地生产复杂花型的织物。

2　基本织造过程

机织物由纵向的经纱和横向的纬纱交织而成。机织结构很多,但就织造过程而言,机织物经以下步骤形成:送经、开口、引纬、打纬和织物牵拉。

为生产机织结构中最简单的平纹织物,来自经轴的经纱越过后梁,绕过分经棒,然后交替穿过两片综框中的综丝,形成两片经纱。送经装置在张力下释放经纱片,开口装置将各综框分别提起或放下。纬纱现在可以通过引纬装置引入,横穿开口,安装在筘座上的钢筘将纬纱打向织口,并压紧它。引入的纬纱这时成为织物的一部分。牵拉辊的连续旋转将刚形成的织物拉离织造位置,然后,织物被卷入卷布辊(再参见图 3.1)。

3 织造准备过程

织造前,必须将经纱卷绕到经轴上,该过程谓之整经。成百上千根经纱通过整经机从落地纱架的纱筒上卷绕到一个大经轴上。在该过程中,为了织造,经纱被涂上混合浆料(比如,淀粉混合物)以增加强度,减少毛羽。然后,经上浆和干燥后的经纱整段或分段平行绕在最终将上织机的经轴上。对于纬纱,必须经络纱以备好用于织造的卷装。无梭织机的纬纱卷绕在圆锥形筒子上,梭织机的纬纱卷绕在纡子上。

4 常见的机织物

机织物的结构多种多样。工艺上,它们可以分为三种:基本组织、变化组织和复合组织。

4.1 基本组织

基本组织有三种,它们是平纹、斜纹和缎纹。

平纹是最常见的机织结构,它由经纬纱分别交替上下交织而成。这类结构的织物广泛用作全棉的和涤/棉的粗平布(床单)和细平布。府绸、有光布、凡立丁、派力司、茧绸、电力纺和塔夫绸都是该结构的常见例子。茧绸和塔夫绸是过去仅用于真丝织物的术语,而今尼龙的塔夫绸(国内常称之"尼丝纺",编者注)和涤纶的茧绸(国内常称之"春亚纺",编者注)等在市场上很常见。

斜纹通过在织物宽度方向纱的交织形成斜向线条或隆起线纹而制得。该结构用于诸如牛仔布、华达呢、哔叽和卡其等结实的产品。它们常用作制服的面料。

缎纹可以分为经面缎纹和纬面缎纹。最常用的是经面缎纹,在经面缎纹中,纬纱在一根经纱上数根经纱下穿过,这样,在织物正面看到的主要是经纱。在纬面缎纹中,纬纱在一根经纱下数根经纱上穿过,这样,在织物正面看到的主要是纬纱。缎纹结构的织物表面光滑,具有光泽。它常用于室内装潢、家庭装饰和时装。缎纹织物常用真丝或化纤长丝织造。

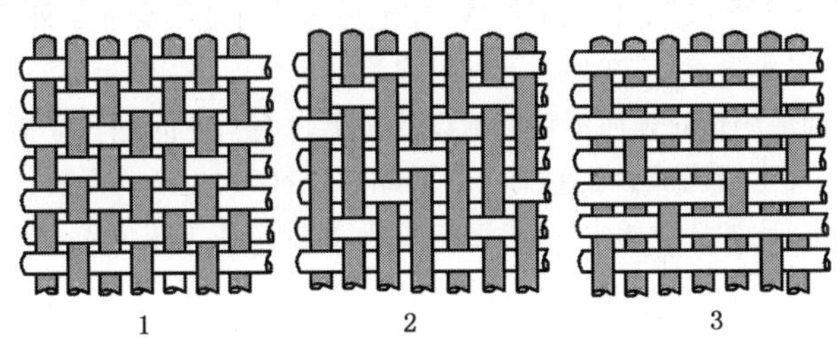

图3.6 基本组织

1—平纹 2—斜纹 3—(纬面)缎纹

4.2 变化组织

顾名思义,变化组织通过基本组织的某种变化派生而得。变化组织有三种:变化平纹组织,比如方平组织和重平组织等;变化斜纹组织,比如菱形斜纹和人字斜纹等;以及变化缎纹组织。

4.3 联合组织或复合组织

将两个或两个以上的上述结构(基本组织或变化组织)联合可以获得联合组织。联合组织的织物广泛用于生产服装和装潢用布,它们一般具有特殊的外观,比如格子效应、蜂窝效应等。如果经纱或纬纱来自两个或两个以上的纱线系统,将获得复合结构。常用的复合结构有二重织物(经二重织物或纬二重织物)、纱罗织物、毛圈织物和灯芯绒等。复合结构的织物广泛用于冬季服装、装饰用布以及工业用纺织物。

5 机织物的质量问题

刚下织机未经进一步处理的织物被称为毛坯布。毛坯布上的任何织疵将影响成品布的质量,尤其是外观质量。

许多外观织疵源于经纱或纬纱的张力不均匀。这种不均匀主要因为机器设置有缺陷。比如,引纬装置故障会使某些纬纱引入时很紧,某些很松,这样纬向条花会显现。如果经纱张力有偏差,可以观察到经向条花。如果整个经纱张力设置得太高,可能会发生“崩纱”(此时,经纱在织造中经常断头)。造成张力不均匀也可能是由于准备纬纱卷装时未充分控制,或,由于整经时张力未设置好。

许多织疵源于机件的问题。综丝或筘片毛口(细小但粗糙的金属突起)会在经纱上留下痕迹,产生综丝擦痕或筘痕。筘痕也可能源于筘隙设置不当。

经纱及/或纬纱的纱疵也会影响织物的质量。纱较高的不匀率或大肚纱会影响织造效率以及织物外观。纱的拉伸强度不佳会导致织物强度不佳。需要注意的是,不同生产批号纱的混用(尤其是合成纤维或它们的混纺纱)在毛坯布上可能不会显示外观问题,但织物染整时可能在织物上显示疵点,或在加工上发生问题。

当然,在织造前或织造时任何不正确地设置机器也会产生问题。穿错综丝或设置织造参数或织机编程时发生错误都会显示疵点。因此,在授权大生产前需要仔细核查样品。此外,织造车间的状况,比如环境的尘埃、湿度的高低,可能造成飞花织入。机器上润滑油过多可能导致坯布上的油污。

第4章　针织和针织物

针织物可以分为纬编的或经编的。不考虑这样的分类,所有针织物都是由纱线线圈串套而成。在纬编针织物中,每个线圈包含两根圈柱、一个针编弧和一个沉降弧。在经编针织物中,一个线圈包含两根圈柱、一个针编弧和一根延展线(见图4.1和4.2)。

横向一排线圈谓之“横列”;纵向一列线圈谓之“纵行”。

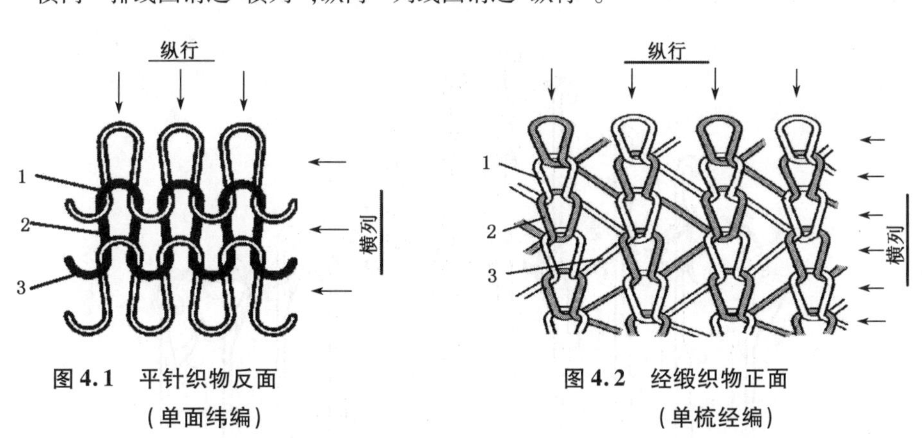

图4.1　平针织物反面
(单面纬编)
1—针编弧　2—圈柱　3—沉降弧

图4.2　经缎织物正面
(单梳经编)
1—开口线圈　2—闭口线圈　3—延展线

单面针织物由一组针编织而成,圈柱和圈弧分别出现在织物的不同一面。仅显示圈柱的那面被称为工艺正面,它较光滑,因为圈柱的配置允许光线有较好的反射。显示圈弧的反面被称为工艺反面,因为光线在圈弧上的漫反射,它看起来较粗糙。使用“工艺”这个术语是因为由该织物生成的最终产品可能使用工艺反面作为产品的实际正面。

如果织物由两组针编织而成,针织物的一面出现圈弧和圈柱,这样的针织物谓之双面针织物。

在纬编编织中,每根纱由相继排列的织针成圈形成一个横列。当这些线圈与相同的织针相继编织而成的线圈串套,生成被称为纵行的一列列线圈。

在经编中,经纱同时成圈形成线圈横列,因此编织前必须整经。

1 成圈过程和成圈机件

为了理解针织物的不同结构和性质,有必要理解织针如何成圈。

1.1 基本成圈过程

线圈由织针和其他成圈机件相互工作而成。织针按通常使用的顺序排,可能是舌针、复合针或钩针。以舌针为例,基本成圈过程或编织动作包含以下步骤:

首先,织针从最低位置升起。在织物的向下牵拉张力作用下,在某些针织机上还借助沉降片喉的握持作用,旧线圈从针钩滑下,打开针舌。当织针继续上升,旧线圈越过针舌勺,到达针杆,完成所谓的退圈阶段。织针现位于它的最高位置(见图4.3中的1至5)。

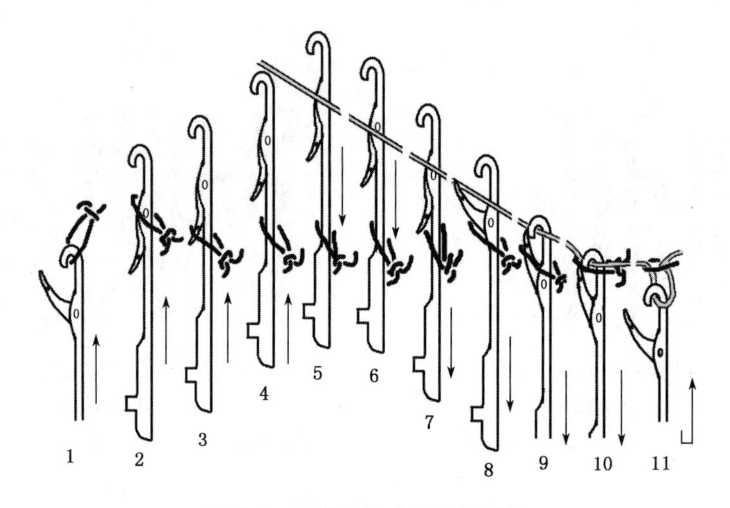

图4.3 舌针纬编成圈过程图示

当织针从它的最高位置下降,新纱引入针钩。织针继续下降,穿过针杆上的线圈,此时,该“旧”线圈关闭针舌,套在针舌上。织针进一步下降,使旧线圈从针头上脱圈,同时将针钩内的纱拉成线圈形状,穿过旧线圈,脱圈形成新线圈。在牵拉力下,新形成的线圈被拉向针背,准备好作为下一线圈的旧线圈(见图4.3中的5至11)。每根织针相继编织,形成一个横列的线圈。

在纬编圆机上,织针插在刻在针筒上的谓之针槽的垂直的槽内。

如果某根织针,握持着旧线圈,不从它的最低位置升起完成退圈,新纱就不能垫入它的针钩。在该纵行,该纱将成为由它形成的横列中的浮线,该针握持的旧线圈将被拉长(见图4.4和4.5,前者显示a针上将形成浮线)。这个动作被称为不编织,所形成的为浮线。

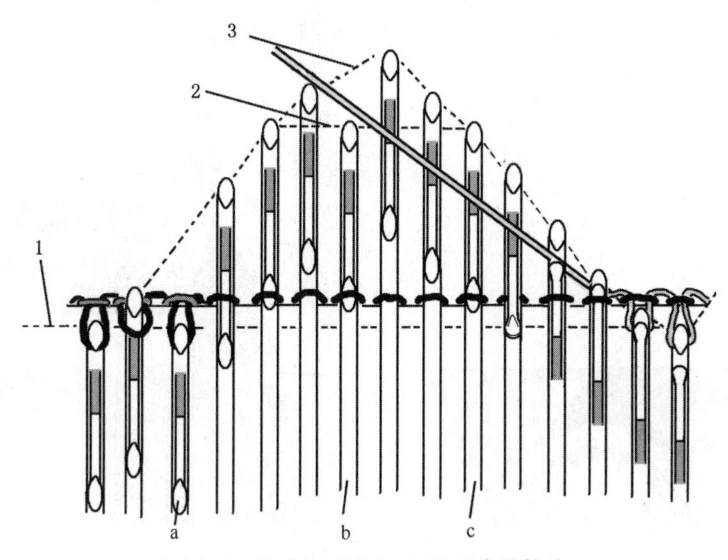

图4.4 织针的不编织、集圈和成圈轨迹

1—不参与编织的织针轨迹 2—集圈针的轨迹 3—参与编织的织针轨迹

如果某根织针仅升起到某一位置,在该位置,旧线圈仍然在针舌上,新垫入的纱将不会从该旧线圈中穿过,只是在其后形成集圈。形成的圈弧就叫集圈圈弧。当织针完成下一个完整的成圈周期,该旧线圈和集圈圈弧再脱圈(见图4.4和4.6,b针和c针上将形成集圈)。这个动作被称为集圈。

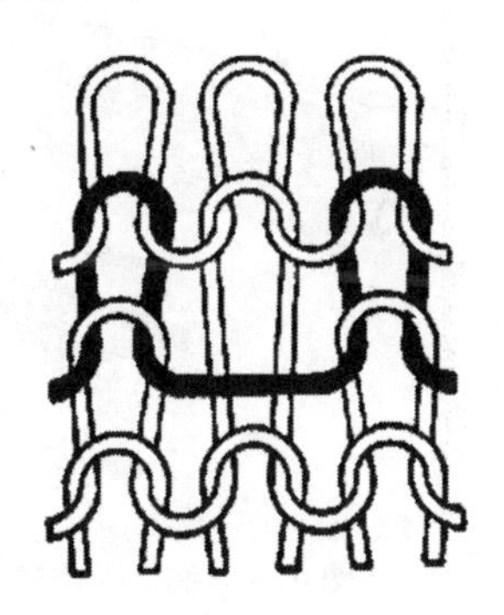

图4.5 因不参与编织而形成的一个单列浮线组织

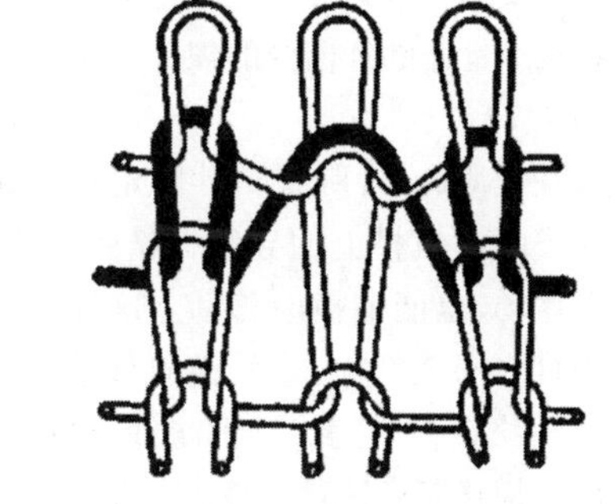

图4.6 一个单列集圈组织

可见,织针的一个上下运动完成一个成圈循环。在纬编中,舌针的上下运动是由织针上突起的针踵和一组诸如起针三角、退圈三角、脱圈三角和弯纱三角等三角组成的三角跑道(见图4.7)的相对运动而产生的。

在圆纬机上,一组这样的三角和一个导纱器构成一路成圈系统。针织机上的成圈系统数决定了针筒一转中可以编织的最大横列数。大多数圆纬机上,三角座固定。针筒的旋转迫使针筒针槽内的织针通过三角跑道上下运动(见图4.8)。

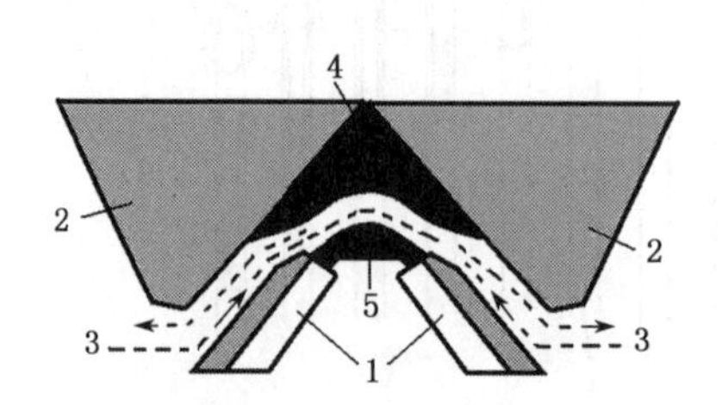

图4.7　横机三角配置

1—起针三角　2—弯纱三角
3—三角跑道　4—压针三角　5—退圈三角

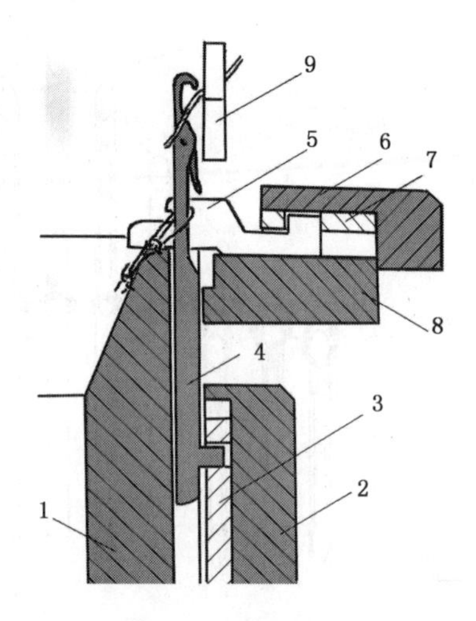

图4.8　圆形单面针织机上的成圈机件

1—针筒　2—三角座　3—三角
4—舌针　5—沉降片　6—沉降片三角环
7—沉降片三角　8—沉降片盘　9—导纱器

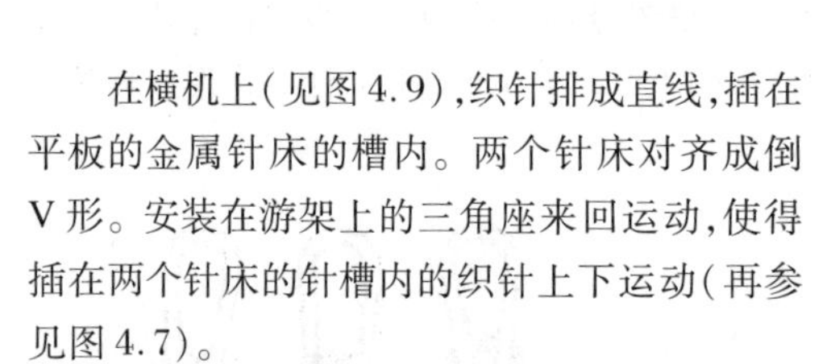

在横机上(见图4.9),织针排成直线,插在平板的金属针床的槽内。两个针床对齐成倒V形。安装在游架上的三角座来回运动,使得插在两个针床的针槽内的织针上下运动(再参见图4.7)。

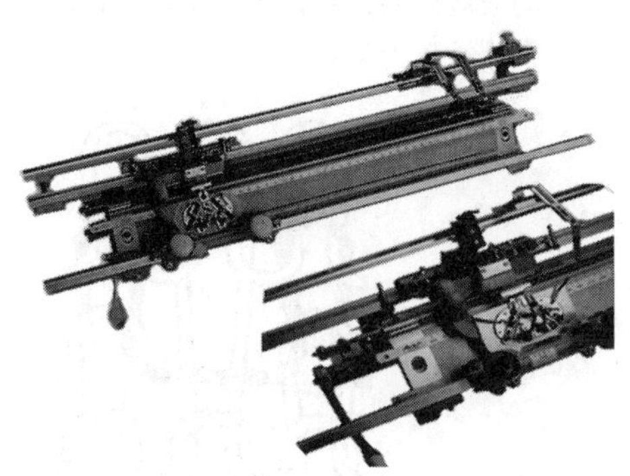

图4.9　手工操作的横机

在经编机上,所有织针固定在针床上,经纱由安装在梳栉上的导纱针携带。梳栉能够前后摆动,也能够横向移动。这样,导纱针通过织针之间至针前,然后横向移动一个针距,将经纱绕在针上完成针前垫纱,而后,从针间摆回并横移完成针背垫纱。因此,针床的一个上下运动以及导纱针的垫纱运动完成一个成圈过程,数百根经纱同时形成一排(横列)的线圈。

1.2　成圈机件

织物通过一系列成圈机件间的联合互动编织而成。成圈机件一般指那些在成圈过程中和纱线接触的机件(再参见图4.8)。如下为这些机件的较详细的说明。

1.2.1　织针

织针是针织机上最基本的机件之一。因为在成圈过程中，在旧线圈的作用下舌针可自动打开或关闭，所以舌针是常使用的织针。从下往上，舌针包含以下部分：针尾、针踵、针杆、针舌、针钩和针头（见图4.10）。安装在针杆槽中的针舌能够绕销子或针杆槽壁定位凹点转动。针舌勺能够确保针钩准确闭合，形成光滑轮廓以便旧线圈越过。织针有不同的粗细，这用机号定义，而机号指机器上单位长度（通常为一英寸）可容纳的织针数。机号越高，单位长度的织针数越多，可编织较细的纱，并且，通常能够生产的织物越薄。

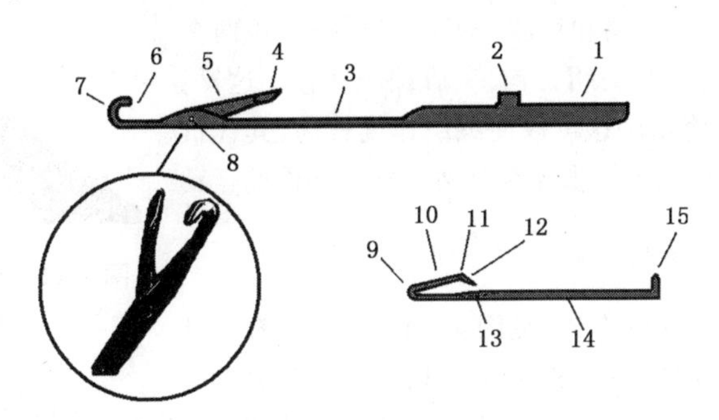

图4.10　舌针和钩针

1—针尾　2—针踵　3—针杆　4—针舌勺　5—针舌　6—针钩　7—针头
8—针舌销　9—针头　10—针钩　11—针鼻　12—针尖　13—针眼槽　14—针杆　15—针踵

针舌工作出问题，几乎不可避免地会在织物上产生疵点。由三角推动的针踵沿三角跑道运动。某些针织圆机的每一成圈系统有一条以上的三角跑道，每条跑道相应对齐不同高度的针踵。这样，相连的织针对同一根纱可以执行一种以上的编织动作。通过设置三角跑道以实施编织、集圈或不编织（通过起针至不同的高度或根本不起针以形成浮线），以及通过以一定顺序配置不同高度针踵的织针，能够生产由成圈、集圈和浮线复合的提花织物。

双面圆形纬编编织需要两组织针的机器，一组在针筒，另一组在针盘。如果针筒针对齐针盘针的针槽壁，反之针盘针对齐针筒针的针槽壁，这样的织针配置谓之"罗纹对针"。这用于罗纹式结构。

另一种对针为"双罗纹对针"。双罗纹对针的机器在针筒和针盘上都使用两组织针。每组针有不同高度的针踵：高蹱或低踵。针筒和针盘的三角座都有两条跑道，一条供高踵针使用，另一条供低踵针使用。针筒高踵针和针盘低踵针对齐，针筒低踵针和针盘高踵针对齐。

钩针（即弹簧针）是另一种传统织针，过去常用它来编织非常细的织物，因为钩针能够做得比其他类型的织针细。钩针具有针头、针钩、针鼻、针尖、针眼槽、针杆和针踵（见图4.10）。钩针的针踵用来将钩针固定在针床上。或者，为了方便，数根针铸成针蜡，然后针蜡再固定在针床上。为了在成圈过程中关闭针钩,使旧线圈套圈,不得不使用压板压针鼻,迫使针尖进入针眼槽,将针钩完全关闭。某些横列可以使用花压板,这样有选择地使针不关闭,旧线圈进入它们的针钩形成集圈。

随着机速的提高,针和压板间的摩擦造成较大发热和噪音,这阻止了机速的进一步提高。结果,大多数原来使用钩针的机器被使用复合针的机器取代。

如今,复合针被广泛使用,特别是在现代经编机上。复合针由两个部件组成:针钩和针芯(即闭口机件)。该两个部件的相对运动,打开及关闭针钩,完成编织动作。这样,针的垂直动程可以降低。这可以使编织速度较高,其代价是要使用较复杂的传动机构来控制这两个独立部件的运动(见图4.11)。

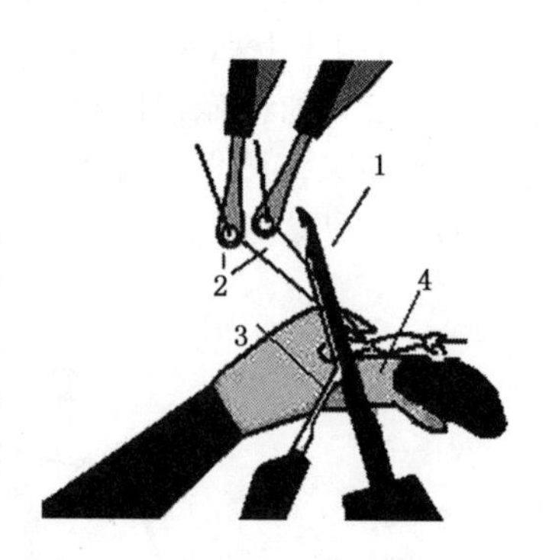

图4.11 带复合针的成圈机件(经编)

1—复合针 2—导纱针
3—针芯 4—沉降片

1.2.2 沉降片

单面针织机上使用沉降片(见图4.8)。沉降片被设计成具有片喉,它在退圈阶段能够握持旧线圈,织针上升时限制旧线圈。沉降片还有片腹,它能够在织针朝最低位置下降时支撑沉降弧,帮助旧线圈脱圈和新线圈形成。

在圆型纬编毛圈机上,专门设计的沉降片鼻能够形成较长的沉降弧,产生毛圈效应。

大多数舌针经编机上,沉降片的形状简单得多,不需片喉或片鼻,因为只用沉降片的下边缘在织针上升时握持织物,所以这样的沉降片谓之握持沉降片。

1.2.3 导纱装置

纬编中,导纱器(见图4.8)送出的纱处于这样的位置,即纱在织针下降时可进入针钩。导纱器的形状因其实施的功能可能随机器种类不同而不同。在基本的舌针圆机上,导纱器成"靴"状,其下部作为护针舌器在旧线圈退离针舌时能够阻止针舌反弹,关闭针钩(使新纱不能垫入)。

如果两根纱在同一路进纱同时编织形成添纱效应,导纱器上将有两个导纱孔,一个用于面纱,另一个用于底纱。两个导纱孔的位置应使纱以不同的垫纱角进入织针,这样,面纱始终显现在织物的正面。

经编中使用导纱针,每根经纱应该穿过单独的导纱针针眼。数根导纱针浇铸成蜡(见图4.12。经编机上的成圈机件通常浇铸成蜡,因为机器上相同的成圈机件始终同步工作),再装上梳栉。梳栉从针背摆向针前,或从针前摆向针背,然后横移,导纱针在针间摆过分别进行针前垫纱(将纱垫入针钩)或进行针背垫纱。梳栉的横移由凸轮、带可更换的不同高度的花板链节的花板轮或电动伺服电机控制。每把梳栉控制一组经纱,它的垫纱运动可以分别控制。经编机的梳栉越多,每一横列的垫纱运动越多(见图4.11和4.13)。

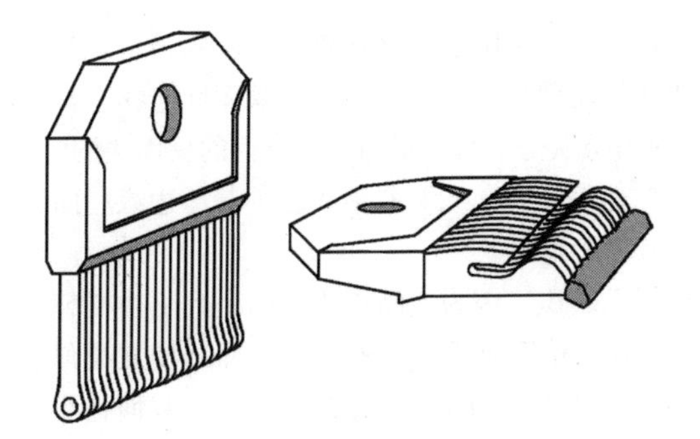

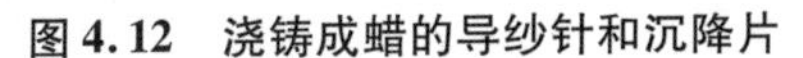

图4.12 浇铸成蜡的导纱针和沉降片

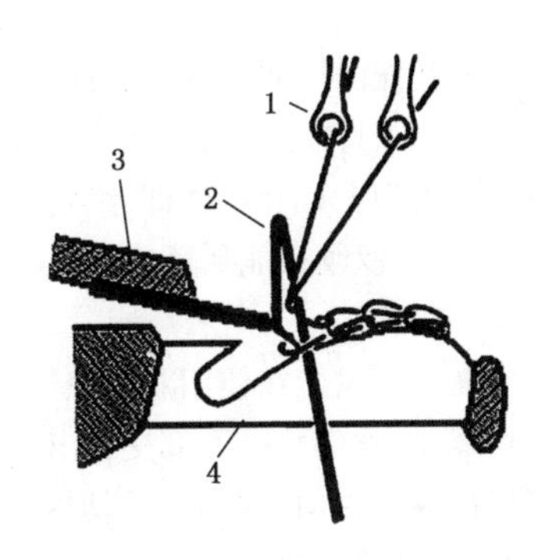

图4.13 带弹簧针(即钩针)的成圈机件(经编)

1—导纱针 2—钩针 3—压板 4—沉降片

2 针织物

由于织物的形成方法不同,纬编织物和经编织物在某些方面有很大不同。

2.1 纬编织物

纬编织物的基本结构为纬平针(单面编织,见图4.1)、罗纹(图4.14)、双反面(见图4.15)和双罗纹(图4.16)。

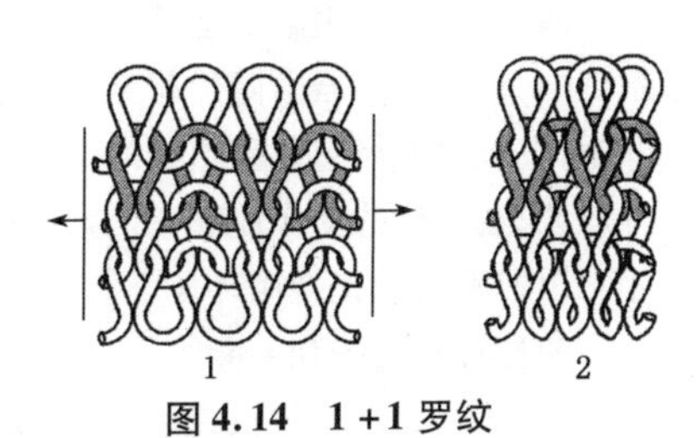

图4.14 1+1罗纹

1—拉伸状态 2—松弛状态

平针织物是最简单但最广泛使用的单面针织物,工艺正面显示圈柱而工艺反面显示圈弧。当裁剪成小块,裁剪的边缘趋向卷边,这在服装生产中可能有些麻烦。平针织物能够沿编织方向或逆编织方向脱散。如果某个线圈的纱线断裂,该纵行的线圈脱散,发生梯脱。平针织物横向或横列方向的延伸性较纵向或纵行方向大。

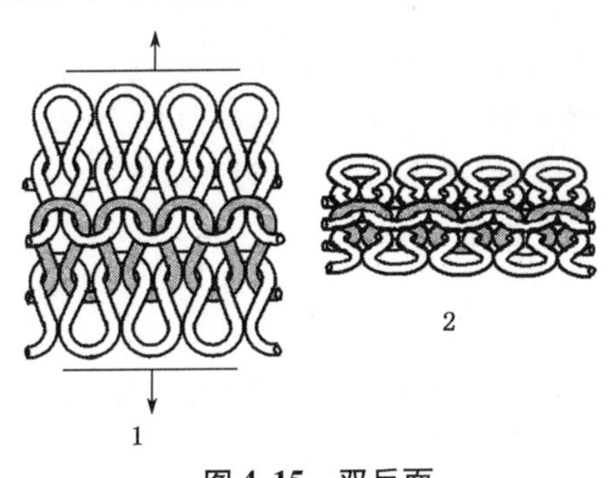

图4.15 双反面

1—拉伸状态 2—松弛状态

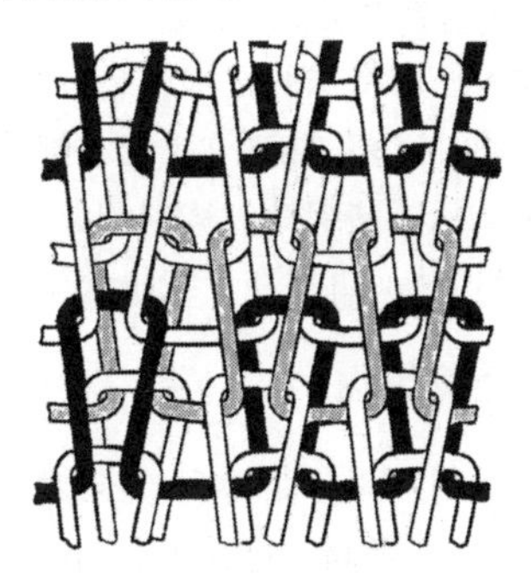

图4.16 双罗纹

罗纹因其横向拉伸力释放后具有很好的回弹性被广泛用于针织服装的袖口或领口。相同正反面线圈纵行间隔配置的罗纹结构是以两面都显示相同外观为特征的双面针织物。常用的罗纹结构为1+1罗纹(即,织物具有交替的一个纵行正面线圈和一个纵行的反面线圈)和2+2罗纹。罗纹只能从最后编织的那端脱散(逆编织方向),只有沿纵行的裁剪边缘趋向卷边。

双反面组织只能在使用双头织针的特殊机器上编织。因为线圈在依次的横列上由交替的针头形成,织物的每一面都交替出现正面线圈横列和反面线圈横列。双反面结构在拉伸后有很好的纵行方向的回复。圆型双反面机有两个重叠的针筒,它们也可以生产双面针织物,常用于双针筒袜类编织。

双罗纹实际是两个1+1罗纹的复合,一个由高踵针形成,另一个由低踵针形成。双罗纹适合做冬季的内衣,因为它具有良好的保暖性。和罗纹一样,双罗纹只能逆编织方向脱散,但是它两个方向的延伸性都较差。

在上述纬编基本结构的基础上,可以有许多花式结构。提花结构是常见的花式结构之一。单面提花结构一般使用机械的或机电的选针机构,通过编织、浮线和集圈的复合而获取。双面提花通常在针筒上借助各种选针机构获取。传统的选针机构使用机械的提花滚筒、中间推片、选针片。

如今,广泛使用电子选针机构,以压电效应为基础的选针系统就是其中一例。在压电式系统中,每一选针器由两片相互黏合的陶瓷盘组成。其选针原理是:当电流通过时,陶瓷盘变弯;电流根据相应花型组织的信号以脉冲方式施加,陶瓷盘微小的动作被选针器中的选针杆放大;选针杆的动作使和它们接触的其他成圈机件偏离,并以此进行选针。电子选针可以生产出非常大的花型循环。

单面衬垫组织和毛圈组织也是常见的花式组织。在前者中,浮线由面纱和地纱使用添纱技术形成的线圈握持;在后者中,毛圈通常借助特殊的沉降片鼻形成。

罗纹和双罗纹变化组织可以在罗纹或双罗纹基础上通过编织、浮线以及集圈横列的复合而获取。它们包括蓬托地罗马双罗纹空气层组织、全畦编、半畦编、法式点纹和瑞士式点纹等。

如果使用特殊的技术或装置,可以获得更多的花式结构。比如,在横机上,通过针床横移或移圈,能够获得波纹或绞花组织。使用调线机构,能够产生多种多样的彩色条纹。通过将针编弧从一些织针上移到相邻的织针上,能够编织纱罗组织。通过移圈技术,可以进行放针或收针,从而直接在针织机上生产无缝成型服装成为可能。

2.2 经编织物

经编机可以分为特利考经编机和拉歇尔经编机。特利考经编机通常机号较高,有2到4个梳栉,机速很高。这种机器的织物牵拉方向几乎和织针平面垂直。

拉歇尔经编机通常梳栉数多得多,由于每把梳栉的横向运动单独由花板轮上各自的花板链或电子推动装置控制,梳栉数多可以编织更复杂的结构。拉歇尔经编机的织物牵拉方向向下,和织针平面成约140°角。

大多数经编机使用单针床,因此只能生产单面针织物。圈柱显示在一面(工艺正面),延展线始终显示在另一面(工艺反面)。有些经编机使用双针床,它们通常用来生产用于汽车内装潢的割绒织物、间隔织物或诸如网袋之类的圆筒织物。

基本经编结构有单梳编链(图4.17)、经平(图4.18)和经缎(见图4.2)。编链结构的特点是一根经纱始终在同一根针上成圈,由其他梳栉衬入的纱沿织物横向将这些线圈连接在一起,形成一块织物。编链具有较低的纵向延伸性,它们经常和其他结构一起使用以获得纵行方向的稳定性。

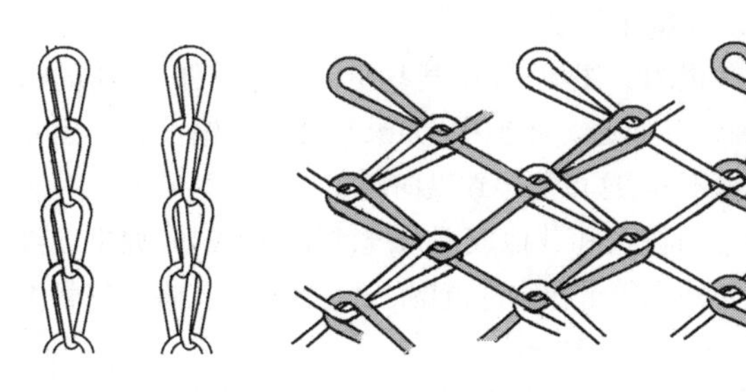

图4.17 编链

图4.18 经平

图4.19 双梳织物(经平绒)

对于经平组织,每根经纱交替在两根相邻的针上成圈。经纱在一根织针上成圈后至少做一个针距的针背垫纱,这样,下一个线圈将在相邻的织针上形成,所有线圈可以连接成织物。

对于经缎结构,携带经纱的导纱针在一个方向的连续数个横列做一个针距的针背垫纱,然后,反向做相同的操作。

如果经平或经缎每次针背垫纱超过一个针距,织物的外观以及延伸性或稳定性会发生变化。一般而言,延展线越长,织物纵行方向的延伸性越大。

在现实中,很少使用单梳经编织物,而使用两把或多把梳栉来编织更稳定的织物,比如经平绒(图4.19)、经绒平、经斜平、经斜编链等等。对于较长延展线浮在反面的结构,反面的延展线可以起绒或刷毛以产生天鹅绒效应。由于前梳栉形成的延展线总是压在其他梳栉形成的延展线上,前梳栉的延展线决定了织物的弹性。它越短,织物结构就越稳定。

通过空穿,可以形成各种双梳栉网眼结构。这是和其他的织物形成方法相比,经编非常大的优点之一。

在网眼结构的基础上,还可以形成各种衬垫花型。这种花型可能通过由各自梳栉控制的衬垫纱的各种各样的轨迹形成,地组织的网眼线圈握持那些衬垫纱。用这样的方法可以编织花边或带花的网眼窗帘布。

如果使用装有贾卡提花机构的机器,待衬入的每根经纱可以单独控制。这样,在编链的基础上,衬垫花型循环可以非常的大。

装上附加的衬纱机构,纱线可以横向、纵向甚至多轴向衬入。由于衬入的纱不需成圈,可以使用粗而刚硬的纱,这使得织物在纱线衬入方向具有非常高的拉伸强度。

3 针织物的质量问题

线圈长度是影响针织物质量的最重要的因素之一,有必要控制线圈长度以确保织物具有良好的品质。许多纬编针织机使用积极喂纱装置以确保线圈长度一致;为了相同的目的,大多数特利考经编机采用积极送经机构。然而,编织较大花型的拉歇尔经编机不得不使用消极送经机构,因为每把梳栉每个横列的送经量相差很大。

从上可见,一个纵行的所有线圈由同一根织针织成,如果某些织针工作不佳,相应的纵行就会出现疵点。织物上纵向条花通常因织针而起。经编中,如果某分段经轴没有正确安装,该经轴上经纱的送经张力将和其他经纱不同,这可能导致纵向条花。

在纬编中,如果某些成圈系统的弯纱三角设置和其他成圈系统的不同,或者,如果纱线张力设置不当,或者任何导纱器垫入的纱的外观质量在容许范围之外,横向条花将在织物上显现。

经编中,通常由于送经张力不匀,或牵拉张力不匀,甚至高速经编机的停机和开机,都会产生横向条花。

除了机器设置不当,成圈机件的缺陷会在织物中产生针洞,或造成花针或漏针。车间温湿度也会造成编织问题,特别在高机号机器上使用合成纤维纱时。

第5章　非织造织物

1　非织造织物的类型

非织造织物绝非是一种新型织物。数个世纪以来,人们就已经知道利用热、水汽和压力使羊毛或动物毛发缠结生产毡布的概念。如今,现代工艺为生产不同用途的非织造织物提供了许多方法。

根据非织造织物的结构,它们可以如下分类。

1.1　以纤维网为底的非织造织物

对于以纤维网为底的非织造织物,首先应该通过梳理成网、湿法成网或气流成网等工艺制备纤维网,然后,通过机械的、化学的、加热或加溶剂手段,使纤维网中的纤维固结在一起形成织物。制备纤维网时,纤维可以多向排列或随机排列,以求各个方向上的稳定性和强度;或它们可以排得更整齐,通常平行于加工方向,以确保纵向的稳定性和强度;或它们也可以沿机器宽度方向斜向排列。还可以加工成复合纤维网,由2到3层不同纤维、或不同定向排列的、或不同成网方法形成的纤维网相互叠合铺放,然后结合,形成具有所需性质的结构。

1.1.1　纤维黏合型

对于黏合法形成的非织造织物,用长丝或者短纤成网后,施加化学剂或热来黏合纤维。

聚丙烯酸酯类的黏合剂是常用的化学剂之一。它可以以粉末或泡沫状施加,或将纤维网在黏合剂中浸透来黏合(见图5.1)。然后,纤维网通过一个热滚筒或通过热风烘炉,将化学剂固化;并且,如有必要,将纤维网烘干。用这种方法形成的非织造织物具有良好的透气性,但手感有些硬。它们的典型用途为墙布或一次性用品。

热黏合是将纤维网黏合的另一种方法。应该使用聚酰胺之类的热熔纤维。通常将纤维网从加热的辊子间通过或绕过这些辊子来加热(见图5.2),某些辊子可以经雕刻,使黏合的纤维网表面轧上凹凸花纹。加热加压使得纤维网中的纤维变软并粘连,一旦冷却,纤维结构内形成黏合。用这种方法形成的非织造织物一般蓬松,具有较好的滤透性、透气性和回弹性。它们可以用作被子或冬装的填料、过滤布或簇绒地毯的基布。

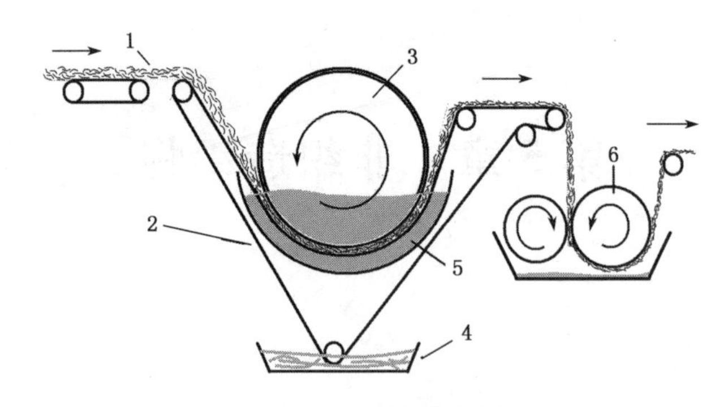

图 5.1　施加黏合剂示例

1—纤维网　2—循环输网帘　3—圆网滚筒　4—网帘清洗槽　5—浸渍槽　6—轧辊

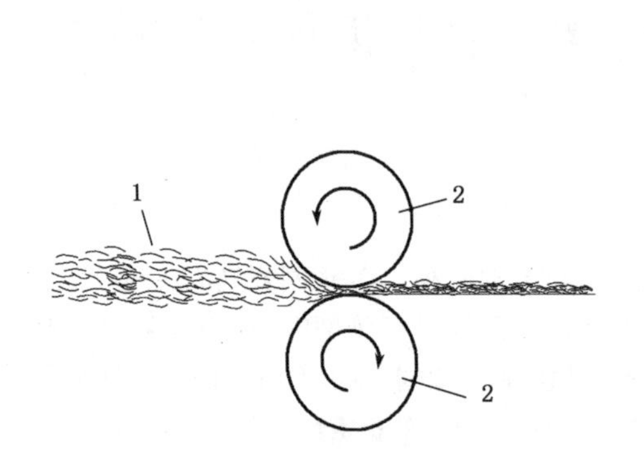

图 5.2　热黏合图示

1—纤维网　2—加热的辊子

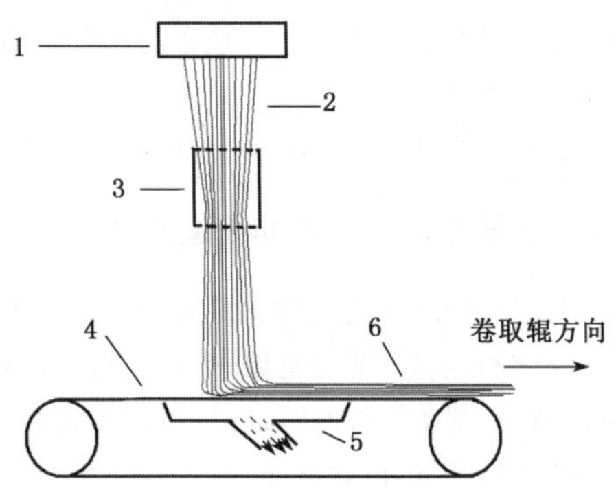

图 5.3　纺黏法纤维层的制备

1—喷丝头　2—冷却区　3—牵伸装置
4—凝网帘　5—吸风装置　6—纤维网

纺黏非织造织物也属此类。当纺丝液从喷丝板挤压出来时，利用静电和高压气流，使纤维随机铺放成纤维薄层（见图5.3），经热滚筒热定形或通过针刺法加固后，制成非织造织物。纺黏织物用于农业，也可用作地毯的底布。

1.1.2　纤维缠结型

这种类型的非织造织物通过诸如针刺和水刺之类的机械方法形成。

在针刺方法中，数千根倒钩针穿透纤维网，迫使纤维从纤维网的一面穿透纤维网并使之缠结固定，形成非织造织物（见图 5.4）。

针刺法，即所谓的针刺成毡法，可以用于多种类型的纤维，特别是羊毛。针刺织物广泛用作土工布、过滤材料、人造革和拼接式地毯的底布。

水刺法的工作原理相似，但它使用细股的高压的水射流而不是倒钩针（见图 5.5）。水射

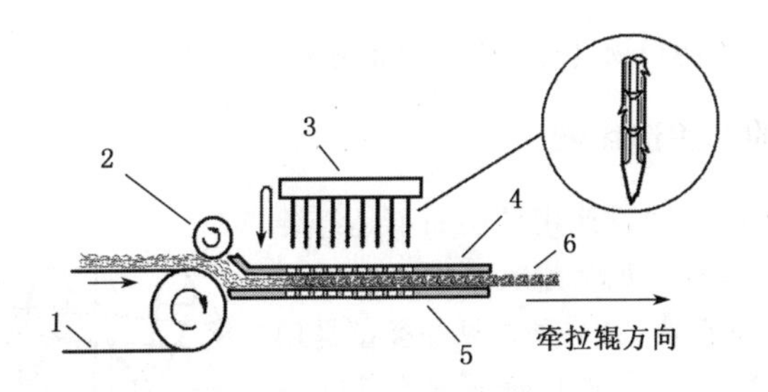

图5.4 针刺示意图

1—输网帘 2—压网辊 3—倒钩针(刺针) 4—压网板 5—托网板 6—针刺织物

流穿透用干法或湿法形成的纤维网,使纤维缠结。当遇到输网帘或支撑纤维网的滚筒时,水射流反弹,这导致纤维网中的纤维进一步缠结。用这种方法生产的织物的英文通常又称为“spun-lace”织物。这类织物的特点是柔软且悬垂性好。

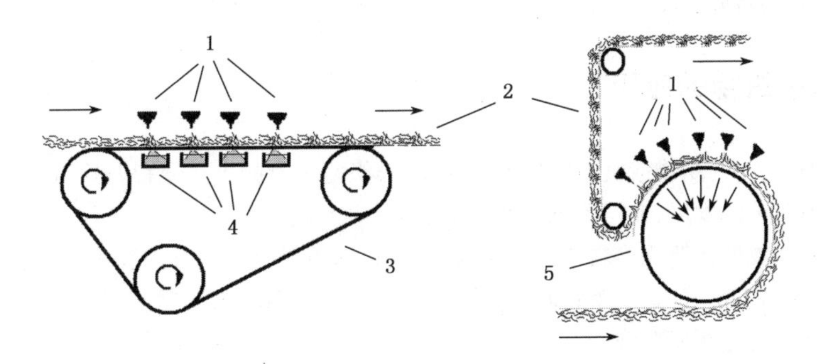

图5.5 水刺法示意图

1—水针 2—纤维网 3—输网帘 4—吸水箱 5—带吸水装置的圆网滚筒

1.1.3 缝编型

纤维网还可以用线圈固结。图5.6为一缝编机示例。它们按与经编机相同的原理工作,织针(通常采用复合针及其针芯)穿透纤维网,和沉降片与导纱针一起工作,形成单梳栉结构以固结纤维网。缝编织物在某种程度上类似针刺毡布,但其线圈结构可以控制,使织物具

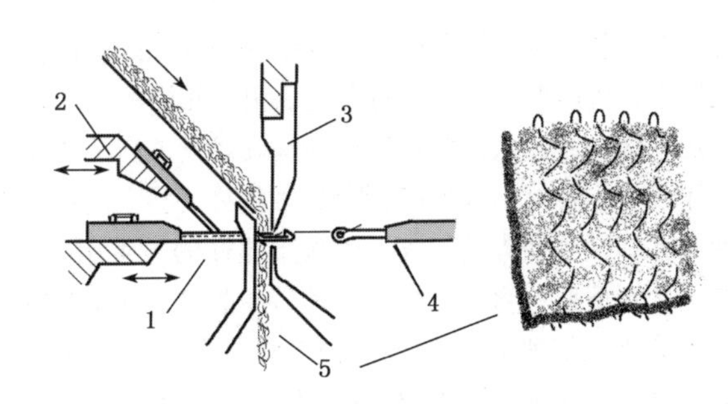

图5.6 缝编示意图

1—复合针针钩 2—针芯 3—沉降片 4—导纱针 5—缝编织物

有一定的方向强度，却不使织物具有不恰当的僵硬感。

1.2 以纱线为底的非织造织物

以纱线为底的非织造织物使用纱层而不是纤维网为底。这类织物的一种常见类型就是用纵向的编链来固结横向的纱层而形成，另一种常见的类型是用经平线圈来增强经纬两向的纱层而形成织物（图5.7）。后一种类型比前一种类型具有较好的稳定性和强度。

可以看到，这类非织造物其实和像多轴向经编织物那样的衬垫经编织物没有明确的差异。如今，为了获得很好的拉伸和冲击性能，某些纤维混合型复合织物就是使用了多轴向经编工艺。

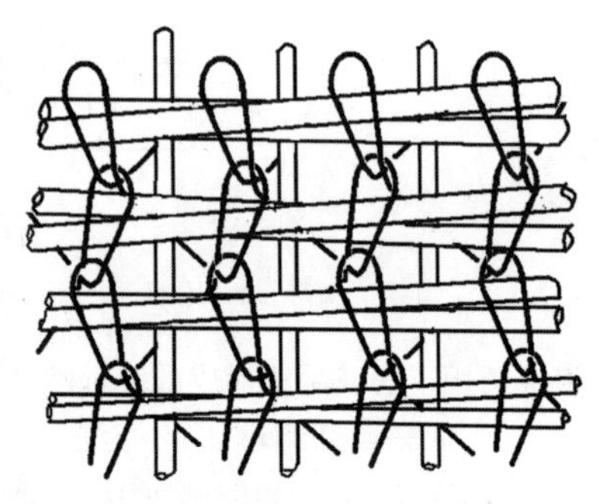

图5.7 以纱线为底的非织造织物

2 非织造织物的应用

可以看到，与针织物或机织物的生产相比，非织造织物的生产流程要短得多，产量要大得多。随着工艺的发展，非织造织物的应用正越来越广泛。

非织造织物的传统用途之一是作为服装辅料，如填料、肩衬、衬头底布。由于它们可做得蓬松，容纳很多空气，这赋予它们保暖性；由于它们可以和服装里料一起裁剪和缝合，使得它们在冬装生产中很受欢迎。

非织造织物的另一个传统用途是用作人造革或地毯的底布。这可以降低生产成本，而同时生产出和机织底布的产品类似的品质。它们在室内装潢和汽车车内装潢中也有相似的用途，比如用作踏脚垫、座垫、墙纸和墙衬。

由于透气性，非织造材料已广泛用作过滤材料。由于纤维科学、非织造织物的成网工艺和织物制造工艺的巨大的发展，非织造织物的过滤器现在不仅广泛用于家用及工业领域，而且广泛用于各种医疗或手术支持系统。由于相同的原因，并由于其相对低的成本，非织造织物现在还在楼房、道路、铁路和河堤的基础建设中广泛用作土工布。

由于具有非常好的吸附性，非织造材料广泛用于生产诸如干纸巾或湿纸巾之类的擦拭纸以及婴儿和大小便失禁的成人用尿布那样的卫生用品。除了考虑吸附性和强度外，非织造织物的生产商们一直在致力于开发易冲散且能够生物降解的擦拭纸。

通过适当的整理工艺，某些水刺织物的强度、透气性、悬垂性和洗涤方面的机械性能可以做得和某些机织或针织布料一样，因此可以用作服装用织物，从而降低服装成本。

在医学领域，非织造织物还用于生产一次性口罩和医生、护士的手术衣。除此之外，非织造的敷料、绷带在医院中正日益受到欢迎。它们在浸渍抗菌剂后可以大大降低感染。

此外,非织造织物的应用已经拓展到军事及高科技领域,它们现已用作飞机和航天飞船壳体以及防弹服的基材。

随着非织造织物的发展和改进,它们的市场在继续成长。因而,它们的用途可能将更广泛。

第6章 染 整

染整是纺织物生产中的重要工序,因为它们赋予最终产品颜色、外观和手感。这类工序取决于使用的设备、纱和织物的组成材料和结构。染整可以在纺织物生产的不同阶段进行。

棉或毛之类的天然纤维可以在纺纱前染色,用这样的方法生产的纱谓之色纺纱。在纺制化学纤维时,染料可以添加到纺丝液中或甚至添加到聚合物切片中。采用这样的方法可生产纺前染色纱。对于色织织物,纱线需要在织造或针织前染色。染色机被设计成可染卷绕较松的绞装或卷绕成一定卷装的纱筒。这样的机器分别被称为绞纱染色机和卷装纱染色机。

整理工序也可以对成衣实施。比如,使用石洗和酵素洗之类的多种方法洗水的牛仔服目前非常流行。对于某些类型的针织服装,还可以成衣染色,以避免服装产生色差。

不过,在大多数情况下,染整对织物进行。先机织或针织成布,然后毛坯布在前处理后染色,及/或印花,并进行化学或机械整理。

1 预处理

为了在染整中获得“可预见的和可复制的”结果,某些前处理是必要的。取决于加工工序,织物可以一匹一匹地处理或一批一批地处理;或为了能连续加工,用可以方便拆除、便于后加工的链缝缝子将不同批次的织物拼接加长。

1.1 烧毛

烧毛是将织物表面的绒毛烧去以避免染色不匀或印花斑渍。一般来说,棉机织毛坯布需要在其他预处理前烧毛。有多种烧毛机,比如平板烧毛机、圆筒烧毛机和燃气烧毛机。平板烧毛机是最简单、最早的类型。待烧毛的布高速通过一块或两块加热的铜板,去除绒毛但不使布烧焦。圆筒烧毛机不使用铜板而使用加热的钢辊,加热能够得到更好的控制。燃气烧毛机现今最常用,燃气烧毛机中织物通过燃气火口烧去表面绒毛。火口的数量和位置以及火焰的长度可以调整以获得最佳效果。

1.2 退浆

如第3章所述,机织中使用的经纱,特别是棉经纱,上浆(通常用淀粉)一般是必要的,以减少纱的毛羽,使纱的强度增加,使其能够承受织造中的张力。然而,留在织物上的浆料可阻止化学剂或染料与织物中的纤维接触。因此,浆料必须在煮练前去除。

将浆料从织物上去除的工序叫退浆。可以采用酶退浆、碱退浆或酸退浆。在酶退浆中,织物用热水浸轧使浆料膨胀,然后浸轧酶液。堆放2至4小时后,用热水洗布。酶退浆需要较少的时间,对织物的损伤较小。但如果使用的不是小麦淀粉而是化学浆料,酶可能无法去除浆料。这时,广泛使用的退浆方法是碱退浆。织物浸渍在烧碱的稀溶液中,并在浸渍槽内堆放2至12小时,然后水洗。如在此之后,织物再经稀硫酸处理,能够获得更好的效果。

由于针织纱不需浆纱,针织物不需退浆。

1.3 煮练

对于天然纤维的毛坯布,纤维上的杂质难以避免。以棉为例,棉花中可能有蜡质、果胶、植物性和矿物性的物质。这些杂质可能使原料纤维呈淡黄色,手感粗糙。纤维中的蜡质、杂质和织物中的油污点可能会影响染色效果。

此外,为了使短纤纱柔软、光滑,具有较低的摩擦系数,以便络纱或针织,上蜡和上油可能是必要的。对于合成长丝,特别是那些用于经编的,在整经中应该使用表面活性剂和抗静电剂,它们通常为专门配制的乳化油。否则,那些长丝可能携带静电,严重干扰编织或织造。

所有杂质、油和蜡必须在染整前去除,煮练可以在很大程度上达到此目的。棉毛坯布精练的最常用方法之一是煮布锅煮练,棉布均匀地堆放在密封的煮布锅内,锅内沸腾的碱液在压力下循环。另一种常用的煮练方法是连续汽蒸,煮练在连续配置的设备中进行,它们一般包括轧液机、J型箱和平洗机。

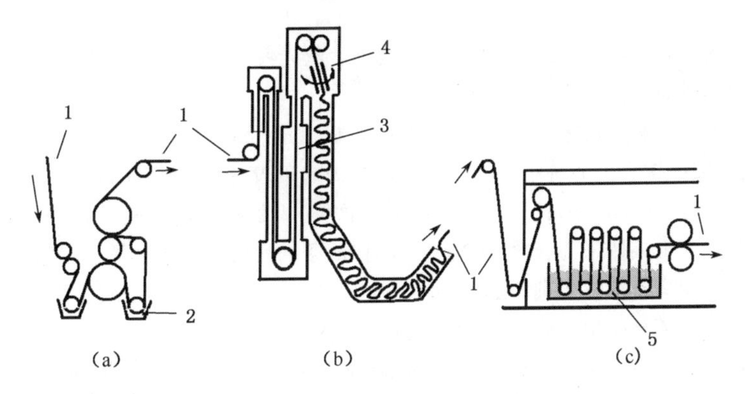

图6.1 连续式汽蒸精练工序设备

(a) 轧液机 (b) J型箱 (c) 平洗机

1—织物 2—碱槽 3—蒸汽加热器 4—甩布器 5—水洗槽

从图6.1可见，碱液通过轧液机施加到织物上，然后，织物被输入J型箱。箱内，饱和蒸汽经蒸汽加热器注入，以后织物均匀堆放。一个多小时后，织物送往平洗机。

1.4 漂白

虽然，大多数棉、麻布中的杂质能够在精练后去除，但织物中仍然留有天然颜色。对于待染成浅色或作为印花底布的织物，有必要漂白以去除内在颜色。

漂白剂实际就是一种氧化剂，通常使用的是以下的漂白剂：

次氯酸钠（也可以使用次氯酸钙）可能是常用的漂白剂。次氯酸钠漂白一般在碱性条件下进行，因为在中性或酸性条件下，次氯酸钠会严重分解，并且纤维素纤维会剧烈氧化，使之成为氧化纤维素。此外，铁、镍和铜类金属及其化合物是次氯酸钠分解的非常好的催化剂，因此，在漂白过程中不能使用这些材料制造的设备。

过氧化氢是一种上乘的漂白剂。用过氧化氢漂白有许多优点。比如，漂白后的织物有良好的白度和稳定的结构，织物强度损伤也比用次氯酸钠漂白小，还有可能将退浆、精练和漂白组合成一个工序。过氧化氢漂白一般在弱碱溶液中进行，应该使用硅酸钠或三乙醇胺之类的稳定剂，以克服上述金属及其化合物造成的催化作用。

亚氯酸钠是另一种漂白剂，它能够赋予织物良好的白度，对纤维的损伤较小，并且也适合连续加工。亚氯酸钠漂白须在酸性条件下进行，不过，随着亚氯酸钠的分解，会释放出二氧化氯气体，它对人体健康有害，并对许多金属、塑料和橡胶产生严重腐蚀，因此一般使用金属钛制造漂白设备，并且将不得不采用必要的措施防止有害气体。所有这些使得这种漂白方法比较昂贵。

1.5 丝光

丝光是在张力下对棉、麻织物的一种碱处理。丝光使棉纤维内腔膨胀，使得它们更有光泽，给予它们丝状外观，并使它们均匀上染。

在丝光中必须对织物施加张力，这是在拉幅机上进行的。织物经过两台轧液机先轧烧碱，然后用安装在拉幅机的链条上的夹子夹住织物的布边，织物随链条向前运动。该两根链条的配置使得链条的间距随链条前移而增加。这样，织物在烧碱影响下的收缩和夹布链条的发散式配置产生夹持张力，拉伸织物。该两根链条的速度可以单独调节，使经纱和纬纱正确对位。织物在离开拉幅机前，通过向织物喷水，然后真空吸去丝光液和洗涤液来进行水洗，这样，当织物离开拉幅机后，它将不会趋于收缩。

丝光可能在漂白前进行，以增强漂白棉府绸和棉印花布之类的织物的白度。如染厚织物，最好在漂白后丝光，以防止染色布上的皱痕。

1.6 热定形

对于大多数合成纤维及其混纺的织物，热定形可能是必要的。在热定形中，织物将拉伸或压缩成所想要的尺寸，然后热定形。热定形可以去除织物上的皱痕，提高织物的热稳定性、机械强度和手感。此外，它会改善织物的染色性能。

湿热定形和干热定形都在使用。在湿热定形中，水被用作溶胀剂以改善定形效果。因此，100%锦纶或其混纺织物可以用湿热定形。然而，100%涤纶或其混纺织物由于涤纶的疏水性采用干热定形。

干热定形机是带有织物输入链条的平幅拉幅机。热定形时，织物通常超喂到针排上，针排握持织物的布边。针排安装在两根链条上，链条向热风烘房移动。两根链条的间距可调整以获得所想要的织物宽度，链条的速度也可以调整以控制热定形时间。约170℃至210℃为适合织物保形和去除皱痕的最佳热定形温度。当织物离开热风烘房后，应该经过一个冷却区，在该处，织物的表面温度降至50℃或50℃以下。适当的织物超喂和横向张力很重要，因为它们影响织物的热稳定性和机械强度。

热定形也可能作为合成纤维织物染整的最后工序。

2 织物染色

织物染色的目的是均匀地赋予织物单种颜色，同时具有良好的色牢度。为了以合理的成本取得最佳的染色效果，染料的类型、助剂、可使用的染色机、加工流程以及诸如加工时间和温度、织物和染液的重量比、染液浓度等工艺参数应该仔细选定。

2.1 常用染料

对于某些诸如纤维素纤维和蛋白质纤维之类的纤维，许多染料可以被使用。然而，对于合成纤维及其织物，染料的选择是有限的。通常用于纺织物的染料如下：

1）直接染料。直接染料是水溶性阴离子染料，它对于纤维素纤维和蛋白质纤维，通过氢键和范德华力，有直接上染性。用直接染料染色时，必须使用软水，否则染料与水中的钙离子或镁离子发生沉淀。当用直接染料染棉织物时，可以使用氯化钠作为促染剂。对于黏胶织物，需要使用匀染剂。可以使用阳离子固色剂改善色牢度。

2）活性染料。活性染料也是水溶性染料。活性染料的活性基团可以和纤维素纤维的羟基很好结合或和蛋白质纤维的氨基以共价键形式很好结合。活性染料染色的织物色彩鲜明，水洗色牢度和摩擦色牢度良好。

3）还原染料。还原染料有可溶性的和不可溶性的。如使用后者，染料先要通过在碱性水溶液中，将其羰基还原成带羟基的隐色体化合物，变成可溶性的，然后对纤维染色，并且氧化回复成原来的不可溶解的形式。可溶性还原染料实际上是含硫酸酯钠盐的隐色体染料，在染纤维素纤维或蛋白质纤维前可先溶于水，对织物染色后，在酸浴中氧化成不溶性形式，然后在纤维上固色。

4）硫化染料。硫化染料为含硫化学剂中衍生的一种染料。它大多用于纤维素纤维。

它有较好的耐洗性,但耐日照性较差。它是最便宜的染料之一,但硫化作用会使纤维素纤维的强度下降。

5）酸性染料。酸性染料是一种阴离子染料。大多数酸性染料是芳香族的磺酸钠盐。它用于诸如羊毛和真丝之类的蛋白质纤维与聚酰胺以及奥纶 44 之类的改性腈纶的染色。

6）分散染料。分散染料是非水溶性的、非离子型的染料。它是随化纤的发展而发展的。早期的分散染料是为纤维素醋酯纤维而研制的。现在,分散染料广泛用于涤纶之类的疏水型的合成纤维染色。分散染料由于分子结构非常简单、非常小而上色,它对于疏水型纤维有直接上染性,使用时需要高压和高温。

7）阳离子染料。阳离子染料又被称为碱性染料。它可以在水溶液中离解以产生带正电荷的色素离子。它用于腈纶和某些如达科纶 64 之类的改性聚酯纤维的染色。

除了染料外,为了获得良好的染色效果,只要适当,就得添加许多其他的助剂。常用助剂有:

- 润湿剂,它用来促使加工液迅速渗透到纤维中;
- 柔软剂,它用来克服织物的粗糙,在染色或其他处理中加入;
- 分散剂,它可以将染料粒子粉碎成更小的粒子以便于它们穿透到纤维中;
- 染色载体,它使疏水性纤维膨胀并使它们具有更多孔隙来吸取染料;
- 多价螯合剂,它和金属杂质结合使它们相对于使用的染料成惰性;
- 固色剂,它使织物内染料颜色牢固。

此外,还可能使用洗涤剂、匀染剂、抗静电剂、消泡剂、防腐剂、防蛀剂、乳化剂和媒染剂等。某些助剂也会在印花或整理工序中使用。所用助剂取决于待染织物的类型、所用染料的类型和所涉及的化学反应的类型。

2.2 织物染色机

染色机主要根据待染织物的类型和所需要的染色效果来选择。

1）卷染机。卷染机适用于纤维素纤维的织物。卷染机有一个装有染液的染槽。染槽上方为一对卷布辊,它们可以顺时针方向或逆时针方向交替转动。槽底有导布辊。待染的织物卷绕在其中的一个卷布辊上,退绕至另一辊上,然后再反向卷绕。导布辊确保织物浸入染浴。槽底配置的蒸汽管能将染液加热到所需温度(图 6.2)。

2）绞盘式绳染机。如果织物不可遭受任何纵向张力,可以使用绞盘式绳染机。机器的染槽中装有染液,染色中的织物浸在染液内。绞盘的转动使织物保持无张力状态下的运动(图 6.3)。

3）轧染机。轧染是最常用的染色方法之一。轧染中,织物可以连续加工。辊式轧染系统有不同的类型,一般它由一台二辊或三辊的轧液机、一条红外线加热通道和一些反应室组成。织物在轧液机上浸轧,再由红外线辐射器加热,然后进入反应室。通过注入蒸汽,反应室的温度保持在一个所需要的水平。反应室后面是一系列进行皂洗和水洗的平洗机。因为轧染对正在染色的织物会造成较大的张力,它不能适用于诸如针织物和弹力织物之类延伸性很好的织物。

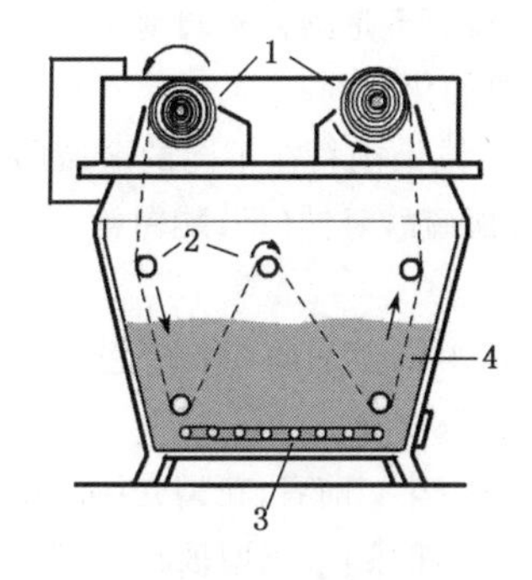

图6.2 卷染机

1—卷布辊 2—导布辊 3—蒸汽管 4—染液

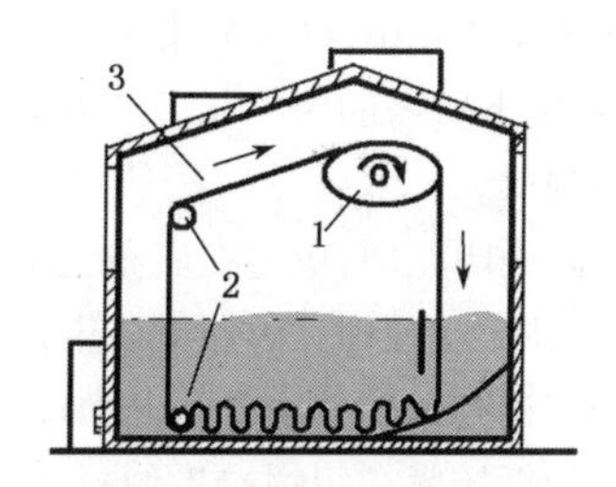

图6.3 绞盘式绳染机

1—绞盘 2—导布辊 3—织物

4）高温高压染色机。一般而言,涤纶之类的合成纤维织物应该在高温高压下染色。这样,染料能够快速而有效地渗透到纤维中。这类机器有多种,比如高温高压绳染机、高温高压溢流染色机和高温高压喷射染色机。

在高温高压溢流染色机中,染色中的织物由染液流推动,因此几乎没有张力施加在织物上,这能够获得较好的染色效果(图6.4)。在高温高压喷射染色机中,喷射装置的作用能够使得染料渗透到纤维内。根据上述两者的工作原理,人们设计了高温高压溢流喷射染色机,它对于弹力合成纤维织物的染色非常有用。

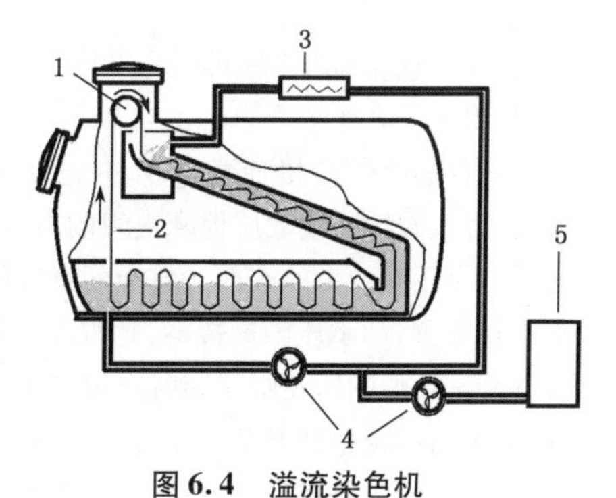

图6.4 溢流染色机

1—导辊 2—织物 3—加热器
4—泵 5—染液补液箱

5）其他设备。为了完成染色工序,有必要使用其他一些设备,比如,用于洗去浮色或进行中和处理的平洗机或绳洗机,用于去除多余水分的离心或真空脱水机,以及将织物烘干的圆网烘燥机或短环烘燥机等。

3 印花

3.1 印花方法

在工艺上,存在数种印花方法,比如直接印花、拔染印花和防染印花等。

在直接印花中,先要制备印花色浆。海藻酸盐或淀粉之类的糊料需要和染料以及其他诸如润湿剂和固色剂之类的必要化学剂按所需要的比例混合,然后根据所需花型将它们印在白色的底布上。对于合成纤维织物,可能使用颜料而不是染料来调制印花色浆,这样的印花色浆可能会包含颜料、黏着剂、乳化糊和其他必要的化学剂。

在拔染印花中,底布先染上所需的底色,然后用拔染糊料在不同的区域印花,拔除或漂白底色,留下所需的白色花型。拔染剂通常用甲醛次硫酸氢钠(即,雕白粉。编者注)之类的还原剂制得。

在防染印花中,应先在底布上施用阻止染色的物质,然后将织物染色。织物染色后,将防染剂去除,花型即在曾印上防染剂的区域显现。

印花还有一些其他类型,比如,转移印花和植绒印花。前者,花型先印在纸上,然后将带有花型的纸压紧在织物或T恤衫之类的服装上,当施加热时,花型被转移到织物或服装上。后者在黏着剂的帮助下将短绒材料按花型印在织物上,通常采用静电植绒印花。

3.2 印花设备

可以通过滚筒印花、筛网印花或近年来的喷墨印花设备印花。

3.2.1 滚筒印花

一台滚筒印花机通常包括一个大型的中心承压滚筒,它覆盖着橡胶或数层棉麻交织的布,使承压滚筒具有光滑且有压缩弹力的表面。数个刻有待印花型的铜辊绕承压辊配置,和承压滚筒接触,每辊印一种颜色。主动旋转的刻有花型的印花辊旋转时也带动了各自的给浆辊转动,后者将印花色浆从它的给浆盘中带给刻有花型的印花辊。一把锋利的被称为色浆刮刀的钢刀将印花辊上多余的色浆刮除;另一把被称为刮绒刀的刀将印花辊黏上的绒毛和污物刮除。待印花的布和毛坯衬布一起从印花辊和承压滚筒间输入,衬布是为了防止色浆渗透织物而弄污承压滚筒的表面(图6.5)。

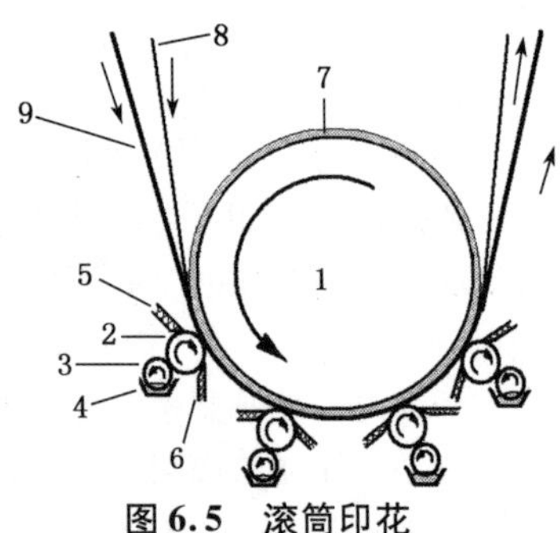

图6.5 滚筒印花

1—承压滚筒 2—印花辊 3—给浆辊 4—给浆盘 5—色浆刮刀 6—刮绒刀 7—橡胶层 8—衬布 9—织物

滚筒印花的产量非常高但刻花印花辊的制备昂贵,这使得滚筒印花实际上仅适用于大批量生产。此外,印花辊的直径限制了花型的大小。

3.2.2 筛网印花

从另一方面说,筛网印花适合较小的订单,并且特别适用于弹力织物的印花。在筛网印花中,先要根据待印花型制备机织网孔布的印花筛网,每个筛网对应一个颜色。在筛网上色浆不得渗透的地方涂有不溶性的薄膜,留下其余网眼孔隙透空,使得色浆可以穿透。通过迫使适量的色浆穿过网眼图案到达筛网下的织物上而完成印花。筛网的制备通过先涂上感光胶,覆上花型的负片,然后将其曝光,使之固定为涂在筛网上的不可溶的薄膜。将涂层上未固化的区域洗去,使筛网上该处留下透空的网眼孔隙。平板式筛网印花是传统的筛网印花,但对于较大的产量要求,圆筒式筛网印花方式也很常用。

3.2.3 喷墨印花

可见,即使许多印花厂已经广泛使用计算机辅助设计系统来辅助花型准备工作,滚筒印花或筛网印花的准备工作仍耗时费钱。必须对待印花型进行分析,以确定将涉及哪些颜色,然后为每一种颜色制备花型负片,并将它转移到印花辊或筛网上。筛网印花的大生产中,不管是圆筒还是平板的,需要经常更换和清洗筛网,这也很费时费力。

为了满足当今市场的快速反应和小批量生产的要求,喷墨印花工艺的使用日益增多。

纺织品喷墨印花使用了和纸张印刷相似的工艺。用 CAD 系统产生的花型的数码信息直接发送到喷墨印花机(更常见的是称之为数码喷墨印花机,用它印花的纺织品可以被称为数码纺织品)并印在织物上。和传统印花工艺相比,该工序简单,并且由于工序的自动化,需要的时间和技术较少,此外,造成的污染也较小。

一般来说,有两种纺织品喷墨印花的基本原理。一种为连续喷墨式(CIJ),另一种为“按需即滴”式。前者,由供墨泵产生的非常高的压力(约 300 千帕斯卡)迫使色墨连续流向喷嘴,喷嘴的直径通常约 10 至 100 微米。在压电式谐振器产生的高频振荡下,色墨被打碎成粒子流并以高速从喷嘴喷出。根据花型,计算机向充电电极发送信号,该电极有选择地对墨粒子充电。当穿过偏转电极,未充电的墨粒子将直接喷向回收槽,而充了电的墨粒子偏向而落在织物上,成为印花的一部分。见图 6.6(a)。

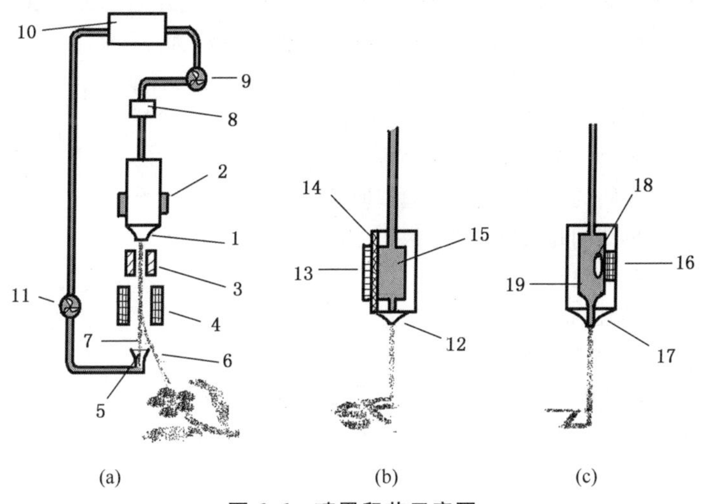

图 6.6 喷墨印花示意图

(a) 连续喷墨法印花 (b) 即滴法印花(机电转换法) (c) 即滴法印花(电热转换法)
1—喷嘴 2—压电式谐振器 3—充电电极 4—偏转电极 5—回收槽 6—已充电墨粒子
7—未充电墨粒子 8—压力调节器 9—供墨泵 10—墨瓶 11—循环泵 12—喷嘴
13—压电装置 14—挠性材料 15—墨室 16—加热器 17—喷嘴 18—气泡 19—墨室

“按需即滴”技术在需要时才供墨，见图6.6(b)。这可以通过机电转换法实施。根据待印的花型，计算机对压电装置发出脉冲信号，该装置变形，通过挠性的中间材料对墨室产生挤压，这种挤压使墨粒子从喷嘴喷出。另一种常用的“按需即滴”技术通过电热转换法进行，见图6.6(c)。在计算机信号下，加热器在墨室中产生气泡。气泡的膨胀力造成墨粒子的喷出。

“按需即滴”式的成本较低，但印花速度较连续喷墨式慢。由于墨粒子连续喷射，连续喷墨式不会发生喷嘴阻塞问题。

喷墨印花通常使用青色、品红、黄色和黑色(CMYK)四种颜色的组合来印制各种颜色的花型。因此应该安装四个打印头，每个对应一种颜色。然而，有些印花机安有两组各八个打印头，因此理论上可以使用多达16种颜色的色墨来印花。喷墨印花机的印花分辨率可达720dpi×720dpi(每英寸点数。编者注)。可以用喷墨印花机印花的织物范围从棉、丝、毛之类的天然纤维到涤纶、锦纶之类的合成纤维，因此需要多种色墨来满足需要。这些色墨包括了活性染料的、酸性染料的、分散染料的甚至颜料型的。

喷墨印花机除了在织物上印花，还可以在T恤衫、圆领长袖运动衫、开领短袖针织衫、婴儿衫、围裙和毛巾上印花。

4 整理

一般而言，织物的整理是通过机械的或者化学的或者两者结合的方法赋予织物所需性能的一种工序。整理能够赋予或提高最终织物的光滑性、绒毛状、悬垂性、光泽、亮度或抗皱性等。某些诸如预缩处理之类的整理工序能够改善织物的最终品质。

对织物采用机械处理的整理工序有许多，例如：

- 在纬编针织物的工艺反面的衬垫纱上刷毛生产绒布；
- 将经编针织物的长延展线拉毛起绒形成毛绒效应；
- 对毛巾布的线圈起绒剪绒产生天鹅绒效应；
- 将织物用滚筒热轧给予它光泽；
- 采用电光整理以获得云纹绸效应；
- 在织物上轧上凹凸花型。

许多粗纺毛织物也需要缩绒、起绒和拉毛之类的机械整理。

还有不少对织物施加化学处理的整理方法，比如，防皱整理、柔软整理、拒水整理、仿桃皮绒整理以及赋予织物抗静电、防腐以及防蛀性能的整理。

某些整理工序同时涉及机械和化学处理。涂层、砂洗和石洗都是很好的例子。

正如本章开头所提及的，整理工序还可能在服装上进行，如今，很多休闲的牛仔服在成衣后进行砂洗、石洗或酵素洗。

整理可以大大改善织物或服装的外观和品质，这就是为什么它是纺织品制造中的关键工序之一，并始终作为许多研究的主题。

5 染整质量问题

成品织物上的疵病可能源于操作失误、机器故障或使用材料的疵病以及不正确的工艺参数。如果操作者不严格按工艺计划中规定的步骤操作，或机器部件失灵，或浴比设置不妥等，被称为“listing”或“tailing”的外观疵病以及条花、色斑或染色不匀等外观疵病可能会发生。所谓的“listing”指染好的织物的布边上产生色差的染色疵病；所谓的“tailing”指的是染好的匹头布的匹头、匹尾产生色差的染色疵病。在某些现代染色机上，染色剂和化学剂的配料根据程序预设的时间和升温曲线自动完成，这大大减少了由于操作者过失导致的染色疵病。

成品织物的内在质量一般应该根据标准测试。比如对色牢度目测评估并且评级。对于光照色牢度，一级最差，八级最好；其他色牢度，比如机械洗涤色牢度、摩擦色牢度、耐气候色牢度和耐汗渍色牢度，一级最差，五级最好。

某些疵病源于毛坯布本身。比如，正如前面章节所提及的，如果不同生产批次的化纤经纱织入同一匹布，如果各批次纱间的上色率不同，纵条色花可能会显现。

除了色牢度，成品织物的其他性质要测量，以检查它们是否在容差范围内，是否符合工艺要求。这些性质包括尺寸变化、拉伸强度、撕裂强度、纱线滑移、抗起球性、折皱回复角、拒水性、纱的线密度、每英寸经纱数或纵行数、每英寸纬纱数或横列数、织物单位面积重量等等。需要检测的具体性质或参数取决于买方和供货方间的协议。双方最好在协议中明确应该遵循的相关标准和允许的容差。必须注意，诸如甲醛残留以及农药残留等方面可能在商业协议中未包含，在这种情况下，有关人员必须根据政府的规定严格加以控制。

（为了方便读者对照阅读，本书中文参考译文部分在保证英文部分原意不变的基础上，尽可能直译。读者在把握原意的基础上完全可以按照中文的修辞法润色。）

Index to Words and Phrases
词汇索引

A

alkaline [ˈælkəlaɪn] condition	碱性条件	第6章1.4
alkaline degradation [ˌdegrəˈdeɪʃən]	碱性降解	第1章2
along the knitting direction	沿编织方向	第4章2.1
amino [ˈæmɪnəʊ] group	氨基	第6章2.1
amorphous [əˈmɔːfəs] regions	无定形区	第1章2
angle of crease recovery	折皱回复角	第6章5
animal fibre	动物纤维	第1章1.1
animal gland [glænd] secretion [sɪˈkriːʃən]	动物腺分泌液	第1章1.1
anionic [ˌænaɪˈɒnɪk] dye	阴离子染料	第6章2.1
anti-foaming agent	消泡剂	第6章2.1
antimicrobial [ˌæntɪmaɪˈkrəʊbɪəl] agent	抗菌剂	第5章2
anti-static	抗静电	第6章4
anti-static agent	抗静电剂	第6章2.1
apparent [əˈpærənt] quality	外观品质	第1章3
apron	围裙	第6章3.2.3
aqueous [ˈeɪkwɪəs] solution	水溶液	第6章2.1
armour [ˈɑːmə]	盔甲	第1章1.2
aromatic [ˌærəʊˈmætɪk]	芳族的	第1章1.2
artificial leather	人造革	第5章1.1.2
asbestos [æzˈbestɒs]	石棉	第1章1.1
atlas [ˈætləs]	经缎	第4章
atmospheric [ˌætməsˈferɪk] pressure	大气压	第1章3
automobile [ˌɔːtəˈməʊbɪl] upholstery	汽车内装潢	第4章2.2
automobile interiors [ɪnˈtɪərɪə]	汽车内装潢	第5章2
auxiliary [ɔːgˈzɪljərɪ] agent	助剂	第6章2
auxiliary nozzle	辅助喷嘴	第3章1

B

back rest	后梁	第3章1
backed weaves	二重组织	第3章4.3
balanced rib structure	平衡的罗纹结构（相同数量的正反面	第4章2.1

	线圈纵行交替配置的罗纹结构)	
bandage [ˈbændɪdʒ]	绷带	第5章2
barbed [bɑːbd] needle	刺针,倒钩针	第5章1.1.2
barre [ˈbɑːreɪ]	纬向条花	第4章3
base cloth	底布	第5章1.1.2
basic dye	碱性染料	第6章2.1
basket structure	方平组织	第3章4.2
bast [ˈbæst] fibre	韧皮纤维	第1章1.1
batching [ˈbætʃɪŋ] roller	卷布辊	第3章1
bath-ratio	浴比	第6章5
beard [bɪəd]	(钩针)针钩	第4章1.2.1
beating-up device	打纬装置	第3章1
bed linen	床单	第3章4.1
bend	(钩针)针鼻	第4章1.2.1
bicomponent [ˌbaɪkəmˈpəʊnənt] fibre	双组分纤维	第2章1
biodegradable	能生物降解的	第5章2
biological organism	生物体	第1章2
bleaching	漂白	第6章1.4
bleaching agent	漂白剂	第6章1.4
blended yarn	混纺纱	第2章2
blending [blendɪŋ]	混棉	第2章2.1
blending ratio	混纺比	第2章2.1
blow-in pipe	吹风管	第2章2.2
boiling shrinkage [ˈʃrɪŋkɪdʒ]	沸水收缩率	第1章3
bonded non-woven fabric	黏合法形成的非织造织物	第5章1.1.1
breaking strength	断裂强度	第1章3
breast beam	胸梁	第3章1
brightness	明亮度	第6章2.1
brushed	经刷毛的	第4章2.2
brushing	刷毛	第6章4
bulk yarn	膨体纱	第2章3

C

cationic dye	阳离子染料	第6章2.1
caustic [ˈkɔːstɪk] soda [ˈsəʊdə]	烧碱	第6章1.2
cellulose [ˈseljʊləʊs] fibre	纤维素纤维	第1章1.2
cellulosic [ˌseljʊˈləʊsɪk] fibre	纤维素纤维	第6章1.4
centrifugal [senˈtrɪfjʊgəl] force	离心力	第2章2.1
centrifugal hydroextractor	离心式脱水机	第6章2.2
ceramic [sɪˈræmɪk] fibre	陶瓷纤维	第1章1.2
ceramic plate	陶瓷盘	第4章2.1
chain stitch	编链	第4章2.2
chain stitch	链缝缝子	第6章1
charge electrode [ɪˈlektrəʊd]	充电电极	第6章3.2.3
chemical composition [kɒmpəˈzɪʃən]	化学成分	第1章2
chintz [tʃɪnts]	有光布,轧光布	第3章4.1
chitosan [ˈkaɪtəsæn] fibre	甲壳素纤维	第1章1.2
chlorine [ˈklɔːriːn] dioxide [daɪˈɒksaɪd]	二氧化氯	第6章1.4
chute [ʃuːt]	(斜向)管道	第2章2.1
circular weft knitting machine	圆纬机	第4章1.1
circular weft terry knitting machine	圆型纬编毛圈机	第4章1.2.2
clamp	夹	第3章1
cleaning	清棉	第2章2.1
cleaning doctor blade	色浆刮刀	第6章3.2.1
cleaning trough [ˈtrɔːf]	清洗槽	第5章1.1.1
clearing cam	退圈三角	第4章1.1
clip	剪	第3章1
clour fastness to laundering	机洗色牢度	第6章5
clour fastness to light	光照色牢度	第6章5
clour fastness to perspiration [ˌpɜːspəˈreɪʃən]	耐汗渍色牢度	第6章5
clour fastness to rubbing	摩擦色牢度	第6章5
clour fastness to weathering	耐气候色牢度	第6章5
CMYK	青色、品红、黄色和黑色的缩写	第6章3.2.3
cohesion [kəʊˈhiːʒən] pressure	抱合力	第2章3.2
cohesive [kəʊˈhiːsɪv] force	抱合力	第2章3.2

crease resistant finishing	防皱整理	第 6 章 4
creep property	蠕变性质	第 1 章 2
crimp frequency	卷曲频率	第 1 章 3
crimping	卷曲	第 2 章 1
cropping [ˈkrɒpɪŋ]	剪绒	第 6 章 4
cross section	横截面	第 1 章 2
crystallinity [ˌkrɪstəˈlɪnɪtɪ]	结晶度	第 1 章 2
crystallization [ˈkrɪstəlaɪˈzeɪʃən]	结晶化	第 1 章 2
cuff	袖口	第 4 章 2.1
cuprammonium process	铜氨法加工	第 1 章 1.2
Cupro [ˈkjuːprəʊ] fibre (CUP)	铜氨纤维	第 1 章 1.2
curl	卷边	第 4 章 2.1
cut (*Ame.*)	机号	第 4 章 1.2.1
cut presser	花压板	第 4 章 1.2.1
cuttling [ˈkʌtlɪŋ] device	甩布装置	第 6 章 1.3
cyan [ˈsaɪən]	青色	第 6 章 3.2.3
cylinder [ˈsɪlɪndə]	针筒	第 4 章 1.1
cylinder	锡林(粗梳机大滚筒)	第 2 章 2.1
cylinder under-casing	锡林下壳罩(大漏底)	第 2 章 2.1

D

decitex [desɪˌteks] (dtex)	分特(克斯)	第 2 章 3.1.1
decompose [ˌdiːkəmˈpəʊz]	分解	第 2 章 1
deflection [dɪˈflekʃən] electrode	偏转电极	第 6 章 3.2.3
delustrant	消光剂	第 2 章 1
denier [ˈdenjə(r)]	旦尼尔	第 2 章 3.1.1
denim [ˈdenɪm]	牛仔布	第 3 章 4.1
dent [dent]	筘隙	第 3 章 1
depressing cam	压针三角	第 4 章 1.1
derivative [dɪˈrɪvətɪv] weaves	变化机织组织	第 3 章 4
desizing [diːˈsaɪzɪŋ]	退浆	第 6 章 1.2
desulphuration [ˌdiːsʌlfəˈreɪʃən]	脱硫	第 1 章 3

drawn sliver	熟条	第 2 章 2.1
Drop on Demand（DOD）	按需即滴式	第 6 章 3.2.3
dropped stitch	漏针	第 4 章 3
dry- laid web	干法成网	第 5 章 1.1.2
dry spinning	干法纺丝	第 2 章 1
dull [dʌl] fibre	消光纤维	第 2 章 1
Dupont Company	杜邦公司	第 1 章 1.2
dust filter box	滤尘箱	第 2 章 2.1
dwell and cross timing	停顿和交错的时间设定	第 3 章 1
dye	染色,染料	第 6 章
dye uptake [ˈʌpteɪk]	染料上色率	第 6 章 5
dye vat	染槽	第 6 章 2.2
dye-carrier	染色载体	第 6 章 2.1
dye-fixing agent	固色剂	第 6 章 2.1
dyeing machine	染色机	第 6 章 2
dyeing unevenness [ˈʌnˈiːvənɪs]	染色不匀	第 6 章 5
dyestuff [ˈdaɪstʌf]	染料	第 6 章 2.1
dye-uptake rate	上色率	第 1 章 3

E

elastic recovery	弹性回复	第 1 章 2
electric current	电流	第 4 章 2.1
electric thermal transfer method	电热转换法	第 6 章 3.2.3
electric voltage [ˈvəʊltɪdʒ]	电压	第 4 章 2.1
electrical resistance	电阻	第 2 章 3.3
electro-conductive yarn	导电纱线	第 2 章 3.3
electromechnical [ɪˌlektrəʊmɪˈkænɪkəl] transfer method	机电转换法	第 6 章 3.2.3
electrostatic [ɪˈlektrəʊˈstætɪk] charge	静电荷	第 6 章 1.3
electrostatic flocking	静电植绒印花	第 6 章 3.1
elongation [ˌiːlɒŋˈgeɪʃən] at break	断裂伸长	第 1 章 3
emboss [ɪmˈbɒs]	凹凸纹	第 5 章 1.1.1

F

fibre-dyed yarn	色纺纱	第6章
filament [ˈfɪləmənt]	长丝	第1章1.2
filling	纬纱	第3章1
filling density	纬密	第3章1
filling face satin	纬面缎纹	第3章4.1
filter cloths	过滤布	第5章1.1.1
filtration [fɪlˈtreɪʃən]	过滤	第5章1.1.1
finished fabric	成品织物	第3章5
finishing	整理	第6章
fixative [ˈfɪksətɪv]	固色剂	第6章2.1
flame	火焰	第6章1.1
flame retardant [rɪˈtɑːdənt]	阻燃的	第1章1.2
flat	盖板	第2章2.1
flat screen printing	平板式筛网印花	第6章3.2.3
flat strip [strɪp]	盖板花	第2章2.1
flat stripping brush	盖板刷帚	第2章2.1
flat stripping comb	上斩刀	第2章2.1
flat V-bed knitting machine	横机	第4章1.1
flax [flæks]	亚麻	第1章1.1
fleece fabric	绒布/纬编针织绒布	第6章4
flexibility [ˌfleksəˈbɪlɪtɪ]	韧性	第1章2
float	浮线	第4章1.1
float stitch	浮线组织	第4章1.1
flock printing	植绒印花	第6章3.1
flushable	能冲散的	第5章2
flyer [ˈflaɪə]	锭翼	第2章2.1
folded yarn	股纱	第2章2.3
foundation [faʊnˈdeɪʃən] construction	基础建设	第5章2
frame	综框	第3章1
French double pique [ˈpiːkeɪ]	法式点纹	第4章2.1
friction spinning	摩擦纺纱	第2章2.2
frictional coefficient	摩擦系数	第6章1.3
fringed [frɪndʒd] selvedge [ˈselvɪdʒ]	毛布边	第3章1

G

H

intermediate [ˌɪntəˈmiːdjət] jack	中间推片	第4章2.1
intermesh	串套	第4章
International Organization for Standardization (ISO)	国际标准化组织	第2章3.1
interstice [ɪnˈtɜːstɪs]	细隙	第6章3.2.2

J

jacquard [dʒəˈkɑːd] mechanism	贾卡提花机构	第3章1
J-box	J型箱	第6章1.3
jig dyeing machine	卷染机	第6章2.2

K

Kevlar fibre	凯芙拉纤维(杜邦公司品牌)	第1章1.2
khaki [ˈkɑːkɪ]	卡其	第3章4.1
kier [kɪə]	煮布锅	第6章1.3
knitted fabric	针织物	第2章
knitting	针织	第4章
knitting action	编织动作	第4章1.1
knitting element	成圈机件	第4章1.1
knitwear [ˈnɪtˌweə]	针织服装	第4章2.1
knock over	脱圈	第4章1.1
knocking-over cam	脱圈三角	第4章1.1

L

lace	花边	第4章2.2
laddering	梯脱	第4章2.1
laid-in angle	垫纱角	第4章1.2.3
laid-in design	衬垫花型	第4章2.2
laid-in yarn	衬垫纱	第4章2.2
laminate [ˈlæmɪneɪt]	薄层	第5章1.1.1

land on	套圈	第4章1.1
lap	棉卷,纤维卷	第2章2.1
lap drum	棉卷罗拉	第2章2.1
lap holder	棉卷架,纤维卷架	第2章2.1
lap rod	棉卷扦,纤维卷扦	第2章2.1
latch	(针)舌	第4章1.2.1
latch guard	护针舌器	第4章1.2.3
latch needle	舌针	第4章1.1
latch needle warp knitting machine	舌针经编机	第4章1.2.2
latch spoon	针舌勺	第4章1.2.1
lateral [ˈlætərəl] movement	横移	第4章1.2.3
leading hooked fibre	前弯钩纤维	第2章2.1
lease [liːs] rod	分经棒	第3章1
length based system	定重制	第2章3.1
leno [ˈliːnəʊ] weave	机织纱罗组织	第3章4.3
let-off device	送经装置	第3章1
leuco [ˈljuːkəʊ] dye	隐色体染料	第6章2.1
leuco-compound [ˈljuːkəʊˈkɒmpaʊnd]	隐色体化合物	第6章2.1
leveling [ˈlevəlɪŋ] agent	匀染剂	第6章2.1
licker-in [ˈlɪkəɪn]	刺辊	第2章2.1
licker-in under-casing	刺辊下壳罩(小漏底)	第2章2.1
linear [ˈlɪnɪə] density	线密度	第2章2.1
linear density deviation [ˌdiːvɪˈeɪʃən]	线密度偏差	第2章3
linen [ˈlɪnɪn]	亚麻织物或纱线	第1章1.1
links and links stitch	双反面组织	第4章2.1
lint doctor blade	刮绒刀	第6章3.2.1
liquor [ˈlɪkə]	液、溶液	第6章1.2
liquor ratio	浴比	第6章2
listing	布边色差	第6章5
localized indentation	定位凹点	第4章1.2.1
locknit [ˈlɒknɪt]	经平绒	第4章2.2
loom [luːm]	织机	第3章1
loom beam	织轴	第3章1

loom fly	织入飞花	第3章5
loop	线圈	第4章
loop forming process	成圈过程	第4章1.1
loop length	线圈长度	第4章3
loop strength	钩接强度	第1章3
loop-clearing	退圈	第4章1.1
low-butt needle	低踵针	第4章1.2.1
lumen [ˈljuːmɪn]	内腔	第1章2
lustre [ˈlʌstə]	光泽	第1章2
Lyocell [laɪəˈsel] (CLY)	绿塞尔(纤维)	第1章1.2

M

machine setting	机器设置	第3章5
magenta [məˈdʒentə]	品红	第6章3.2.3
magnesium ion	镁离子	第6章2.1
malfunction [mælˈfʌŋkʃən]	故障	第3章5
mangle [ˈmæŋgl]	轧液机	第6章1.3
man-made fibre	化学纤维	第1章1
mattress [ˈmætrɪs]	垫子,踏脚垫	第5章2
mean and coefficient [ˌkəʊɪˈfɪʃənt] variation	平均值及变异系数	第1章3
mechanical properties	机械性质	第1章2
medical dressings [ˈdresɪŋz]	医用辅料	第5章2
melt spinning	熔体纺丝	第2章1
melting temperature	熔点	第2章1
mercerization [ˌmɜːsəraɪˈzeɪʃən]	丝光	第6章1.5
mercerizing [ˈmɜːsəraɪzɪŋ] liquor	丝光液	第6章1.5
mesh structure	网眼结构	第4章2.2
meta-aramid fibre	间位芳族聚酰胺纤维	第1章1.2
metal fibre	金属纤维	第1章1.2
metric count	公支	第2章3.1.2
micro fibre	超细纤维	第2章1
micron [ˈmaɪkrɒn]	微米	第1章

N

nanoparticle [ˈnænəʊpɑːtɪkl]	纳米粒子	第1章1.2
nanotoxicology [nænətɒksɪˈkɒlədʒɪ]	纳米毒理学	第1章1.2
narrowing	收针	第4章2.1
nap	毛绒,拉毛	第6章1.1
napkin [ˈnæpkɪn]	餐巾,纸巾	第5章2
napping	拉毛	第6章4
natural fibres	天然纤维	第1章1
neck	领口	第4章2.1
needle	织针	第4章
needle bar	针床	第4章1.1
needle felting [ˈfeltɪŋ]	针刺	第5章1.1.2
needle head	针头	第4章1.1
needle loop	针编弧	第4章
needle punching	针刺	第5章1.1.1
needle selection mechanism	选针机构	第4章2.1
needle selector	选针器	第4章2.1
needle space	针距	第4章2.2
needled fabric	针刺织物	第5章1.1.2
negative image	负片	第6章3.2.2
negative let-off mechanism	消极送经装置	第4章3
negative let-off system	消极送经系统	第3章1
net sack	网袋	第4章2.2
net structure	网眼结构	第4章2.2
neutral [ˈnjuːtrəl]	中性的	第6章1.4
neutralization [ˌnjuːtrəlaɪˈzeɪʃen]	中和	第6章2.2
nip roller	轧辊	第2章2.1
Nomex fibre	诺梅克斯纤维(聚间苯二甲酰间苯二胺纤维,杜邦公司品牌)	第1章1.2
non-ionic dye	非离子型染料	第6章2.1
non-knitting	不编织	第4章1.1
non-woven fabric	非织造物	第5章1.1
non-wovens	非织造物	第5章1.2

pad dyer	轧染机	第 6 章 2.2
pad-roll system	辊式轧染系统	第 6 章 2.2
Palace	派力司	第 3 章 4.1
partial or part set threading	部分穿经	第 4 章 2.2
paste [peɪst]	糊料	第 6 章 3.1
pattern chain	花板链	第 4 章 1.2.3
pattern drum	提花滚筒	第 4 章 2.1
pattern link	花板	第 4 章 1.2.3
pattern repeat	花型循环	第 3 章 1
pattern wheel	花板轮	第 4 章 2.2
patterned fabric	带花纹的织物	第 4 章 1.2.1
patterned mesh curtain	带花纹的网眼结构窗帘	第 4 章 2.2
peaching	仿桃皮绒整理	第 6 章 4
pectin ['pektɪn] product	果胶产物	第 6 章 1.3
periodic [pɪərɪ'ɒdɪk]	周期性的	第 2 章 3.3
photogelatin	感光胶	第 6 章 3.2.2
pick density	纬密	第 3 章 1
picking stick	打梭杆	第 3 章 1
piezo [paɪ'iːzəʊ] system	压电系统	第 4 章 2.1
piezoelectric [paɪˌiːzəʊɪ'lektrɪk]	压电的	第 4 章 2.1
piezoelectric vibrator [vaɪ'breɪtə]	压电式谐振器	第 6 章 3.2.3
pigmented ['pɪgməntɪd] ink	颜料型色墨	第 6 章 3.2.3
pile fabric	割绒织物	第 4 章 2.2
pill resistance	抗起球性	第 6 章 5
pillar	圈柱	第 4 章
pin clips	针排	第 6 章 1.6
pirn [pɜːn]	纡子	第 3 章 1
plain knit	平针	第 4 章
plain weave	平纹	第 3 章 4.1
plain weave derivatives	平纹变化组织	第 3 章 4.2
plain woven fabric	平纹织物	第 3 章 2
plaiting [plætɪŋ, pleɪtɪŋ] device	甩布装置	第 6 章 1.3

plant fibre	植物纤维	第1章1.1
plate singer [ˈsɪndʒə]	平板式烧毛机	第6章1.1
plating effect	添纱效应	第4章1.2.3
plating techniques	添纱技术	第4章2.1
plied yarn	股纱	第2章2.3
ply-yarn	股纱	第2章3
polar [ˈpəʊlə] group	极性基团	第1章2
polo shirts	短袖开领针织衫	第6章3.2.3
polyamide [pɒlɪˈæmaɪd] (PA)	锦纶/聚酰胺	第1章1.2
polybutylene [ˌpɒlɪˈbjuːtɪliːn] terephthalate (PBT)	聚对苯二甲酸丁二酯	第1章1.2
polyester [ˈpɒlɪestə] (PES)	涤纶/聚酯	第1章1.1
polyester pongee	春亚纺	第3章4.1
polyester – cotton sheeting	涤/棉粗平布	第3章4.1
polyethylene [ˌpɒlɪˈeθɪliːn] (PE)	聚乙烯	第1章1.2
polyethylene terephthalate (PET)	聚对苯二甲酸乙二酯	第1章1.2
polymer	聚合物,聚合体	第1章1.2
polymer chip	聚合物切片	第2章1
polymerisation [ˈpɒlɪmeraɪz, pəˈlɪ-]	聚合化	第1章2
polymerise [ˈpɒlɪməraɪz, pəˈlɪ-]	聚合	第1章1.2
polypropylene [ˌpɒlɪˈprəʊpɪliːn] (PP)	丙纶/聚丙烯	第1章1.2
polytrimethylene [ˈpɒlɪtraɪˈmeθɪliːn] terephthalate [ˈteˌrefˈθæleɪt] (PTT)	聚对苯二甲酸丙二酯	第1章1.2
polyurethane [ˌpɒlɪˈjʊərɪθeɪn] (PU)	氨纶/聚氨基甲酸酯	第1章1.2
pongee [pɒnˈdʒiː]	茧绸	第3章4.1
Ponte de Roma	蓬托地罗马双罗纹空气层组织	第4章2.1
poplin [ˈpɒplɪn]	府绸	第3章4.1
porosity [pɔːˈrɒsɪtɪ]	透气性、多孔性	第5章1.1.1
positive let-off mechanism	积极送经装置	第4章3
positive yarn feeding device	积极喂纱装置	第4章3
precipitate [prɪˈsɪpɪteɪt]	沉淀	第6章2.1
predictable [prɪˈdɪktəb(ə)l]	可预见的	第6章1
preliminary [prɪˈlɪmɪnərɪ] treatment	预处理	第6章1

pre-programmed	程序预设的	第6章2
pre-shrinkage treatment	预缩处理	第6章4
presser	压板	第4章1.2.1
presser bar	压板床	第4章1.2.3
pressure bowl	承压滚筒	第6章3.2.1
pressure cylinder	承压滚筒	第6章3.2.1
pressure regulator [ˈregjʊleɪtə]	压力调节器	第6章3.2.3
primary weaves	基本机织组织	第3章4
printing	印花	第6章3.1
printing head	打印头	第6章3.2.3
printing paste	印花色浆	第6章3.1
printing roller	印花辊	第6章3.2.1
production lot	生产批次	第3章5
production output	产量	第5章2
production sequence [ˈsiːkwəns]	产生流程	第5章2
projectile [prəˈdʒektaɪl; (*US*)-tl]	片梭	第3章1
projectile loom	片梭织机	第3章1
projectile catching-brake	接梭制梭器	第3章1
projectile conveyer	传梭器	第3章1
projectile feeder	喂梭器	第3章1
projectile lift	提梭器	第3章1
prone [prəʊn] to	易于……的	第1章2
proportion of over-length fibres	倍长率	第1章3
pulsed signal	脉冲信号	第6章3.2.3
purl [pɜːl] stitch	双反面组织	第4章2.1

Q

queenscord	经斜编链	第4章2.2
quilt [kwɪlt]	被子	第5章1.1.1

R

rabbit hair	兔毛	第1章1.1

resultant [rɪˈzʌltənt] linear density	总线密度	第2章3.1.2
retaining [rɪˈteɪnɪŋ] plate	挡板	第2章2.2
reverse locknit	经绒平	第4章2.2
rib	罗纹	第4章2.1
rib derivative structure	罗纹变化组织	第4章2.1
rib gated structure	罗纹对针的结构	第4章1.2.1
rib gating	罗纹对针	第4章1.2.1
ring	钢领	第2章2.1
ring spinning	环锭纺	第2章2.1
rinsing [ˈrɪnsɪŋ]	水洗	第6章2.2
ripple stitch	波纹组织	第4章2.1
ripstop effect	方格效应	第3章4.3
rivet [ˈrɪvɪt]	针舌销	第4章1.2.1
roller printing	滚筒印花	第6章3.2
roller singer	圆筒烧毛机	第6章1.1
roller washing machine	平洗机	第6章1.3
rope washer	绳洗机	第6章2.2
rotary screen drum	圆网滚筒	第5章1.1.1
rotary screen dryer	圆网烘燥机	第6章2.2
rotary screen printing	圆筒式筛网印花	第6章3.2.3
rotor [ˈrəʊtə]	转杯	第2章2.2
rotor spinning	转杯纺纱	第2章2.2
rot-proofing	防腐	第6章4
rot-proofing agent	防腐剂	第6章2.1
roving [ˈrəʊvɪŋ]	粗纱	第2章2.1
roving bobbin [ˈbɒbɪn]	粗纱筒子	第2章2.1
roving frame	粗纱机	第2章2.1

S

S twist	S捻	第2章2.2
sand washing	砂洗	第6章4
sateen [sæˈtiːn]	纬面缎纹	第3章4.1

satin [ˈsætɪn]	缎纹,经面缎纹	第3章4.1
satin derivative weave	缎纹变化组织	第3章4.2
saturated [ˈsætʃəreɪtɪd] steam	饱和蒸汽	第6章1.3
saw cut	针舌槽	第4章1.2.1
scaly [ˈskeɪlɪ] surface	鳞片层表面	第1章2
schreinering [ˈʃraɪnərɪŋ]	电光整理	第6章4
scorching [ˈskɔːtʃɪŋ]	烤焦	第6章1.1
scouring [ˈskaʊrɪŋ]	煮练(精练)	第6章1.3
screen	筛网	第6章3.2.2
screen printing	筛网印花	第6章3.2
seamless garments	无缝成型服装	第4章2.1
seat cushion [ˈkʊʃən]	座垫	第5章2
second passage drawframe sliver	二道熟条	第2章2.1
sectional beam	分段经轴	第4章3
selection jack	选针片	第4章2.1
self-twist (ST) yarn	自捻纱	第2章2.2
self-twist spinning	自捻纺纱	第2章2.2
self-twist, twisted (STT) yarns	加捻自捻纱	第2章2.2
selvedge tuck-in device	布边折入装置	第3章1
semi-dull fibre	半消光纤维	第2章1
sensor	传感器	第3章1
separating plate	隔纱板	第2章2.1
sequestrant [sɪˈkwestrənt]	多价螯合剂	第6章2.1
serge [sɜːdʒ]	哔叽	第3章4.1
shank	(针)尾,(针)杆	第4章1.2.1
shape retention	保型性	第6章1.6
sharkskin [ˈʃɑːkskɪn]	经斜平	第4章2.2
shed [ʃed]	开口,梭口	第3章1
shedding [ˈʃedɪŋ] device	开口装置	第3章1
shell casing	壳体	第5章2
shell fabric	面料	第3章4.1
shirting	细平布,衬衣料	第3章4.1
short loop hanging dryer	短环烘燥机	第6章2.2

shoulder padding	肩衬	第5章2
shuttle	梭子	第3章1
shuttle box	梭箱	第3章1
shuttle weaving	梭织	第3章1
shuttleless loom	无梭织机	第3章1
signal	信号	第3章1
singeing [sɪndʒɪŋ]	烧毛	第6章1.1
singeing machine	烧毛机	第6章1.1
single bar warp knit	单梳经编	第4章
single jersey [ˈdʒɜːzɪ]	单面针织物	第4章
single jersey laid-in stitch	单面衬垫组织	第4章2.1
single knit	单面针织	第4章
single needle bar	单针床	第4章2.2
single weft knit	单面纬编	第4章
singles	单股纱	第2章2.3
singles yarn	单股纱	第2章3
singles yarn	单纱	第2章3.2
sinker	沉降片	第4章1.2
sinker belly	沉降片片腹	第4章1.2.2
sinker cam	沉降片三角	第4章1.2
sinker cam ring	沉降片三角环	第4章1.2
sinker dial	沉降片盘	第4章1.2
sinker lead	沉降片蜡	第4章1.2.3
sinker loop	沉降弧	第4章
sinker nose	沉降片片鼻	第4章1.2.2
sinker throat	沉降片喉	第4章1.1
Sirofil yarn	赛络菲尔纱	第2章2.2
Sirospun yarn	赛络纱	第2章2.2
sized and dried yarns	经上浆并烘干的纱	第3章3
sizing compound	上浆复合浆料	第3章3
sley [sleɪ]	筘座	第3章1
slider	(复合针)针芯	第4章1.2.3
sliding or closing element	针芯	第4章1.2.1

spindle [ˈspɪndl]	锭子,芯轴	第2章2.1
spinneret [ˈspɪnəret]	喷丝头	第2章1
spinning	纺纱或细纱(工艺),纺丝	第2章1
spinning mill	纺纱厂	第2章2.1
spring or bearded needle	弹簧针/钩针	第4章1.2.1
spun yarn	短纤纱	第2章2
spun-bonded fabric	纺黏非织造织物	第5章1.1.1
spun-dyed yarn	纺(前)染(色)纱,纺液着色纱	第6章
spun-lace fabric	水刺非织造织物	第5章1.1.2
squeezing roller	轧辊	第5章1.1.1
stability	稳定性	第5章1.1
stabilizer [ˈsteɪbɪlaɪzə]	稳定剂	第6章1.4
stand creel	落地纱架	第3章3
staple fibre	短纤	第2章
starch mixture	淀粉混合物	第3章3
starch paste	淀粉糊料	第6章3.1
static [ˈstætɪk] inhibitor [inˈhɪbɪtə(r)]	抗静电剂	第6章1.3
stationary [ˈsteɪʃ(ə)nərɪ]	固定的	第3章1
steeping [ˈstiːpɪŋ]	浸渍,浸泡	第6章1.2
stem	(针)杆	第4章1.2.1
stenter [ˈstentə]	拉幅机	第6章1.5
stitch	(针织)组织,线圈	第4章1.1
stitch-bonded fabric	缝编织物	第5章1.1.3
stitch-bonding machine	缝编机	第5章1.1.3
stitching cam	弯纱三角	第4章1.1
stone washing	石洗	第6章
streakiness	条花	第4章3
striper mechanism	调线机构	第4章2.1
strippers	剥离辊	第2章2.1
stripping roller	剥棉罗拉	第2章2.1
sublistatic printing	转移印花	第6章3.1

T

tex [teks]	特克斯	第2章3.1.1
textiles	纺织品	第1章
textured	经变形的	第2章1
textured yarn	变形纱	第2章3
thermo-fusible fibre	热熔性纤维	第5章1.1.1
thread	线	第2章2.3
tip	(钩针)针尖	第4章1.2.1
titanium [taɪˈteɪnjəm, tɪ-]	钛	第6章1.4
titanium dioxide[daɪˈɒksaɪd]	二氧化钛	第2章1
to be circulated [ˈsɜːkjʊleɪtɪd]	被循环	第6章1.3
to be overfed	被超喂	第6章1.6
tolerance [ˈtɒlərəns]	容差	第1章3
trailing hooked fibre	后弯钩纤维	第2章2.1
traveller	钢圈	第2章2.1
triacetate [traɪˈæsɪˌteɪt] (CTA)	三醋酯(纤维)	第1章1.2
trick	针槽	第4章1.1
tricot	经平	第4章2.2
Tricot [ˈtrɪkəʊ] machine	特利考经编机	第4章2.2
tri-ethanolamine [ˌtraɪeθəˈnɒləmiːn]	三乙醇胺	第6章1.4
trumpet [ˈtrʌmpɪt]	喇叭口	第2章2.1
T-shirt	T恤衫,短袖圆领针织衫	第6章3.2.3
tubular [ˈtjuːbjʊlə] product	圆筒产品	第4章2.2
tuck	花针	第4章3
tuck loop	集圈	第4章1.1
tuck stitch	集圈组织	第4章1.1
tuck-in selvedge	折入的布边	第3章1
tucking	集圈	第4章1.1
tufted [ˈtʌftɪd] rug	簇绒地毯	第5章1.1.1
tufts of fibre	纤维簇	第2章2.1
tussah [ˈtʌsə]	柞蚕丝	第1章1.1
twill	斜纹	第3章4.1
twill weave derivative	斜纹变化组织	第3章4.2

twist	加捻	第2章2.1
twist factor	捻系数	第2章3.2
twist level	捻度	第2章3.2
twist unevenness [ˈʌnˈiːvənɪs]	捻度不匀	第2章3
twisting	加捻	第2章1
two or three-roller padder	二辊或三辊式轧液机	第6章2.2

U

ultraviolet [ˌʌltrəˈvaɪəlɪt] (UV) light	紫外线光	第1章2
unconventionally shaped fibre	异形纤维	第2章1
underlap	延展线	第4章
underlap	针背垫纱	第4章1.1
unravel [ʌnˈrævəl]	脱散	第4章2.1
unwind [ˈʌnˈwaɪnd]	退绕	第2章2.1
upholstery [ʌpˈhəʊlstərɪ]	室内装潢	第3章4.1

V

vacuum hydroextractor [ˈhaɪdrəʊɪksˈtræktə]	真空脱水机	第6章2.2
Valitin	凡立丁	第3章4.1
Van der Waals [ˈvæn dɜːˌwɔːlz] force	范德华力	第6章2.1
variabilities [ˌveərɪəˈbɪlɪtɪs] in strength or elongation	断裂强度或伸长的偏差	第1章3
vat dye	还原染料	第6章2.1
vegetable and mineral substances	植物及矿物性物质	第6章1.3
vegetable content	杂草屑	第1章3
vegetable fibre	植物纤维	第1章1.1
velvet [ˈvelvɪt]	天鹅绒	第4章2.2
vertical line	纵向条花	第4章3
viscose [ˈvɪskəʊs] (CV)	黏胶	第1章1.2
viscose process	黏胶法加工	第1章1.2
visual weaving fault	外观织疵	第3章5

vortex [ˈvɔːteks] spinning	涡流纺纱	第2章2.2

W

wadding [ˈwɒdɪŋ]	填料	第5章1.1.1
wale	纵行	第4章
wall covering	墙布	第5章1.1.1
wall lining	墙衬	第5章2
wall paper	墙纸	第5章2
warp	经纱	第3章1
warp "smashing"	经纱崩纱	第3章5
warp "streakiness"	经向条花	第3章5
warp backed weave	经二重组织	第3章4.3
warp beam	经轴	第3章3
warp face satin	经面缎纹	第3章4.1
warp knitting machine	经编机	第4章2.2
warp run-in	经纱送经量	第4章3
warp sheet	经纱片	第3章1
warping	整经	第3章3
washing	水洗	第2章1
washing liquor	洗涤液	第6章1.5
water injector	水针	第5章1.1.2
water repellency [rɪˈpelənsɪ]	拒水性	第6章5
water repellent finishing	拒水整理	第6章4
water-jet loom	喷水织机	第3章1
waxing	上蜡	第6章1.3
weaving	机织,织造	第3章
weaving machine	织机	第3章1
weaving position	编织位置	第3章1
weaving technology	织造工艺	第3章1
web based non-woven fabric	以纤维网为底的非织造织物	第5章1.1
web plate	托网板	第5章1.1.2

worsted yarn	精纺毛(型)纱	第2章2.1
woven fabric	机织物	第2章
wrap [ræp] yarn	包覆纱	第2章3

Y

yarn	纱	第2章
yarn carrier	导纱器	第4章1.1
yarn feeder	导纱器	第4章1.1
yarn guide	导纱器	第3章1
yarn package	纱的卷装	第2章2.1
yarn slippage [ˈslɪpɪdʒ]	纱线滑移	第6章5
yarn spinning	纺纱	第2章2.1
yarn-based non-woven fabric	以纱线为底的非织造织物	第5章1.2
yarn-dyed fabric	色织布	第6章
yarn-guiding hole	导纱孔	第4章1.2.3

Z

Z twist	Z捻	第2章2.2